INTRODUCTORY DIFFERENTIAL AND INTEGRAL CALCULUS

BRIAN K. SALTZER

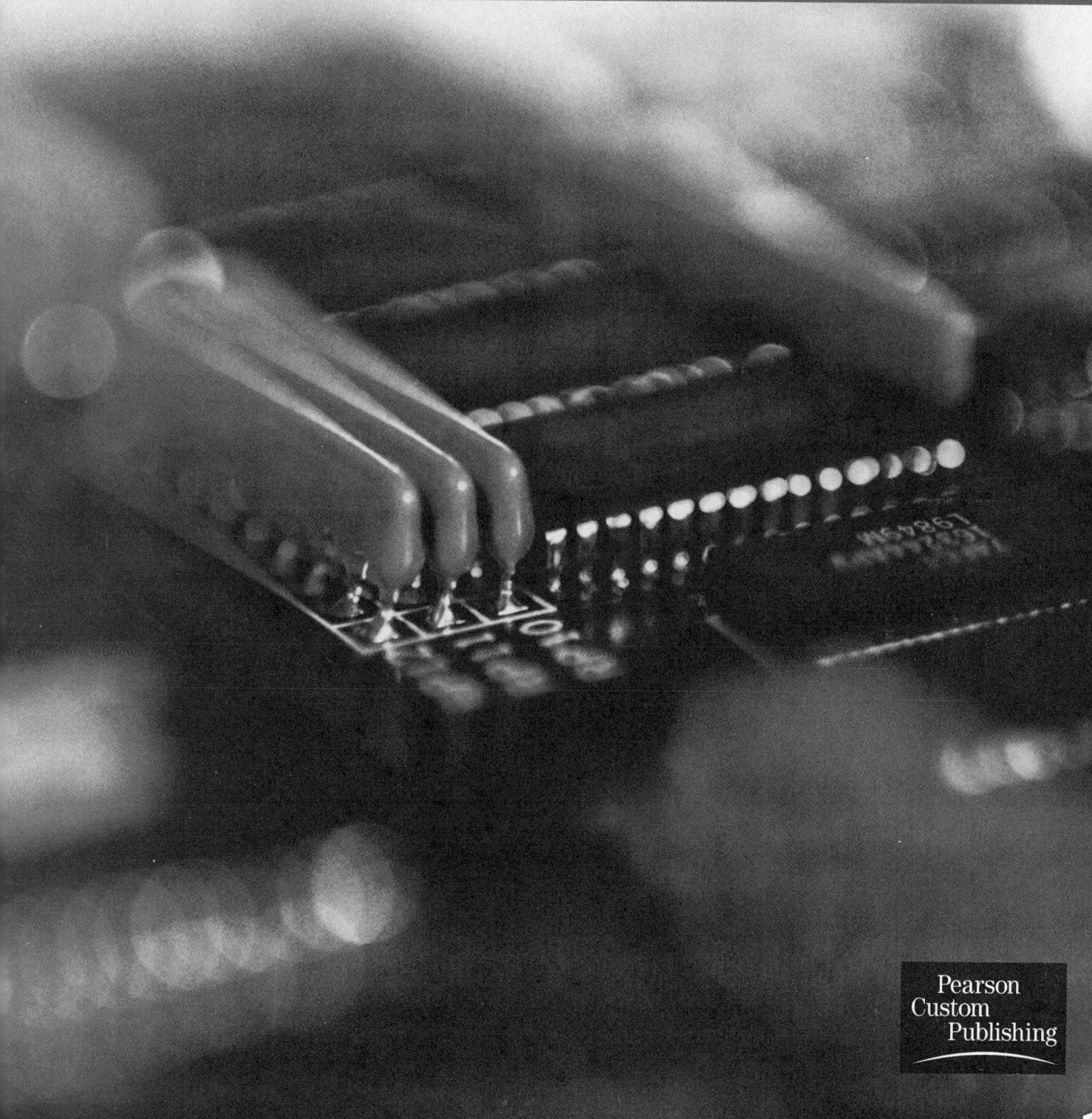

Pearson Custom Publishing

Cover designed by Mary Louise Dorfner.

Copyright © 2002 by Pearson Custom Publishing
All rights reserved.

Permission in writing must be obtained from the publisher before any part of this work may be reproduced or transmitted in any form or by any means, electronic or mechanical, including photocopying and recording, or by any information storage or retrieval system.

Printed in the United States of America

10 9 8 7 6 5 4 3 2 1

This manuscript was supplied camera-ready by the author.

Please visit our web site at www.pearsoncustom.com

ISBN 0-536-63499-8

BA 993732

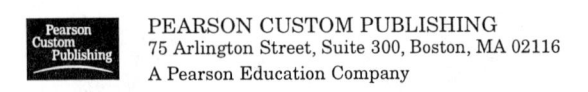

PEARSON CUSTOM PUBLISHING
75 Arlington Street, Suite 300, Boston, MA 02116
A Pearson Education Company

For my mother, Shirley Saltzer, with love and thanks.

Contents

Preface   vii

Differential Calculus

1.	Derivative Preliminaries	3
2.	The Derivative of a Function	23
3.	Rules for Finding Derivatives	37
4.	Differentials I – An Alternative Method of Expressing Derivatives	67
5.	Higher-Order Derivatives	75
6.	Implicit Differentiation	81
7.	Maxima, Minima, and Points of Inflection	87

Differential Calculus Applications   103

Integral Calculus

1.	Integration Preliminaries	115
2.	Finding the Integral of a Function	123
3.	The Method of Substitution	143
4.	Differentials II – The Relationship Between Integral and Differential Calculus	151
5.	Integration by Parts	159
6.	Numerical Integration Techniques	171

Integral Calculus Applications   181

Solutions to Exercises   191
Index   241

Preface

▸ The Purpose of the Book

My intention in writing this book was to write a calculus book that a beginning calculus student could read on his/her own. Unfortunately, when most students are presented with calculus for the first time they are so buried in mathematical formalism that the techniques and applications elude them. Accordingly, I have attempted to write a book that is as "user-friendly" as possible.

The material is presented in a manner as simplified as possible and I make no claim of addressing the deeper mathematical rigor. I am of the opinion that the deeper insights in any area of mathematics come with experience and not through the first presentation. It is only after a student has some working knowledge of the methods and procedures that are indigenous to a particular area of mathematics that the deeper, subtler connections become apparent.

My wish is that this book will show at least one beginning calculus student that the subject is not as frightening as he/she might have believed.

▸ The Structure of the Book

The book is divided into two halves, *Differential Calculus* and *Integral Calculus*. At the end of each of the two sections are application exercises that are intended to illustrate the benefits of learning the topic. These application exercises are drawn from a variety of technical and non-technical subject areas.

To aid the student in learning the various calculus skills, Exercises with Solutions have been included at the end of each section as well as additional Practice Exercises. In a break with mathematical tradition, the solutions to most exercises have been left in an "unsimplified" form so that students can see whether or not they have correctly executed a particular calculus algorithm. For years, I have watched students' frustration with this point. They execute an algorithm to find the derivative, integral, etc., and then try for twenty minutes to prove that their answer is equivalent to the "simplified" one in the back of the book.

▸ Supplements

An Instructors' Resource Manual with solutions to the Practice Exercises is available from the publisher.

Acknowledgements

First, thanks goes to Frank Burrows and the production staff at Pearson Custom Publishing for making the book possible.

Next, I would like to thank the faculty members and students of ITT Technical Institute – Pittsburgh for their time and efforts in reviewing the manuscript and working all of the exercises. As always, it's a privilege to be associated with you.

Considerable thanks goes to Chester Benson for his extremely fast and accurate copyediting and to Ann Heath for her many comments on the manuscript.

Lastly, my largest thanks goes to Eric Stimmel for his excellent artwork and for formatting the text.

Brian K. Saltzer
Pittsburgh, PA

Differential Calculus

- Derivative Preliminaries
    - The Slope of a Line
    - Calculating the Slope of a Line
    - Interpreting the Slope of a Line
    - Dependent and Independent Variables
    - The Limit of a Function

    Exercises with Solutions
    Practice Exercises

- The Derivative of a Function

    Exercises with Solutions
    Practice Exercises

- Rules for Finding Derivatives
    - Rule 1 - Simple Polynomials

    Exercises with Solutions
    Practice Exercises

    - Rule 2 – Functions Involving the Sine and Cosine
    - Rule 3 – Functions Involving the Exponential Function
    - Rule 4 – Functions Involving the Natural Logarithm

    Exercises with Solutions
    Practice Exercises

    - Rule 5 – The Product Rule
    - Rule 6 – The Quotient Rule
    - Rule 7 – The Power Rule

    Exercises with Solutions
    Practice Exercises

- Differentials I – An Alternative Method of Expressing Derivatives

    Exercises with Solutions
    Practice Exercises

- Higher-Order Derivatives

    - Higher-Order Derivatives Using Functional Notation
    - Higher-Order Derivatives Using Differentials

    Exercises with Solutions
    Practice Exercises

- Implicit Differentiation

    Exercises with Solutions
    Practice Exercises

- Maxima, Minima, and Points of Inflection

    - Finding the Extreme Values of a Function Using The First Derivative Test
    - Finding Maxima and Minima Using the Second Derivative Test
    - Absolute Versus Local Maximum and Minimum Values
    - Points of Inflection

    Exercises with Solutions
    Practice Exercises

- Differential Calculus Applications

DIFFERENTIAL CALCULUS

1

Derivative Preliminaries

▶ The Slope of a Line

In order to understand the idea of the derivative of a function, it is imperative that we have a strong grasp of the concept of the slope of a line. Although it may not appear so at the outset of our study, we will see that the derivative of a function is nothing more than an algebraic expression that will give us the slope of a curve at any point. To gain an intuitive feel for the concept of the derivative, let's review how to calculate the slope of a line as well as how to interpret the slope once it has been found.

> **Math Review**
>
> Remember that a set of coordinate axes is nothing more than two number lines that have been placed perpendicular to one another.

Suppose we have a set of coordinate axes as in the following:

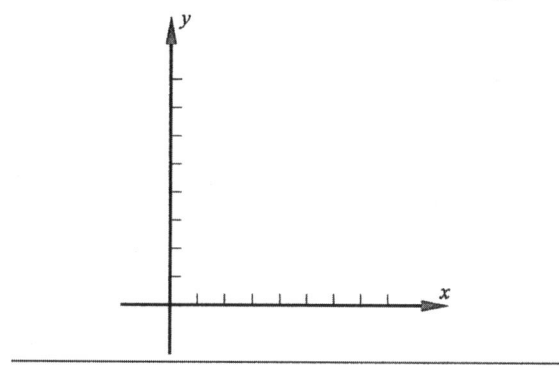

Fig. D.1.1

It is conventional to call the horizontal axis the "x-axis" and the vertical axis the "y-axis."

To place a point on the graph, we must know how far out the x-axis to go and how far up the y-axis to go. For example, if we move 2 units out the x-axis, and 3 units up the y-axis, we arrive at the point (2, 3). Remember that the coordinates of a point are always expressed with the x-coordinate placed first and the y-coordinate placed second.

This is how we plot the point (2, 3):

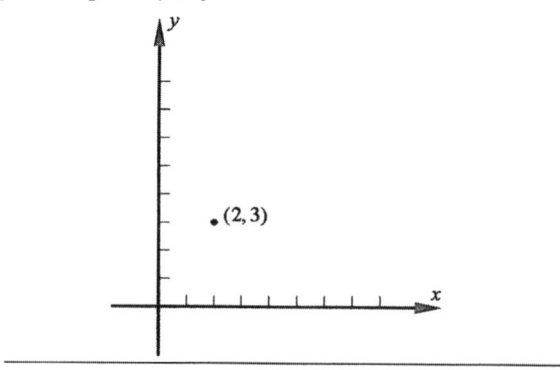

Fig. D.1.2

In order for us to place a line on our graph, we must know two of the points that lie on the line. Suppose we have the following two generic points:

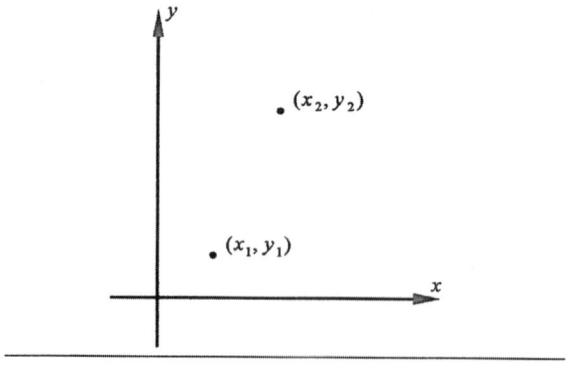

Fig. D.1.3

We can now use the two points to create a line on our graph. Remember that only a single, unique line can be drawn through two points:

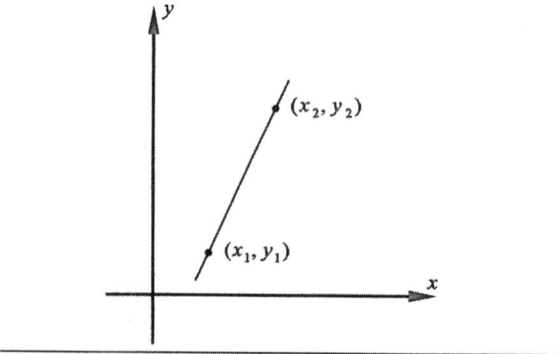

Fig. D.1.4

Using the two points that determined the line, we can now discuss the idea of the *slope* of the line. Our discussion breaks into two distinct parts:

1. Calculating the slope of a line
2. Interpreting the slope of a line

▸ Calculating the Slope of a Line

Look again at our graph with the line and two points.

To calculate the slope of a line, take the difference in the *y*-coordinates and divide it by the difference in the *x*-coordinates.

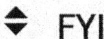 **FYI**

The slope of a line is often referred to as the *rise* over the *run*.

If we write out this slope definition using the coordinates of our two generic points, we get:

$$\text{slope} = \frac{y_2 - y_1}{x_2 - x_1}$$

Let's use our slope equation to calculate the slopes of three different lines.

Example 1
Point 1 = (1, 2)
Point 2 = (3, 5)

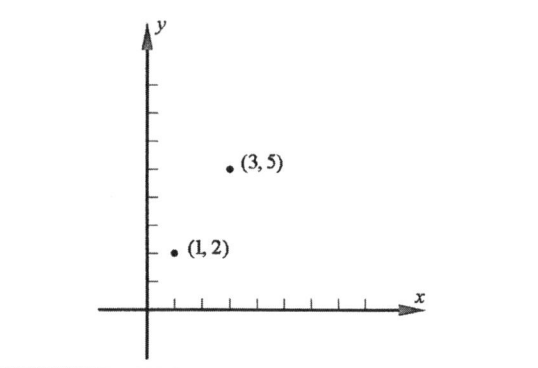

Fig. D.1.5

We now draw the line determined by our two points:

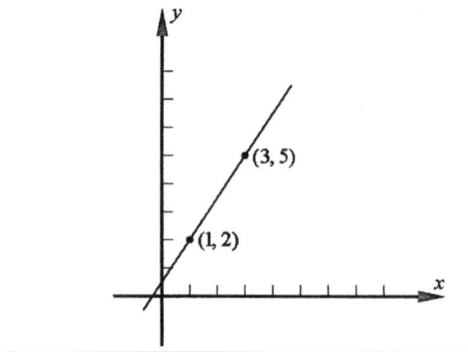

Fig. D.1.6

Next, we insert our two sets of coordinates into the slope formula:

$$\text{slope} = \frac{y_2 - y_1}{x_2 - x_1}$$

$$\text{slope} = \frac{5 - 2}{3 - 1}$$

$$\text{slope} = \frac{3}{2}$$

Example 2

Point 1 = (1, 8)
Point 2 = (6, 2)

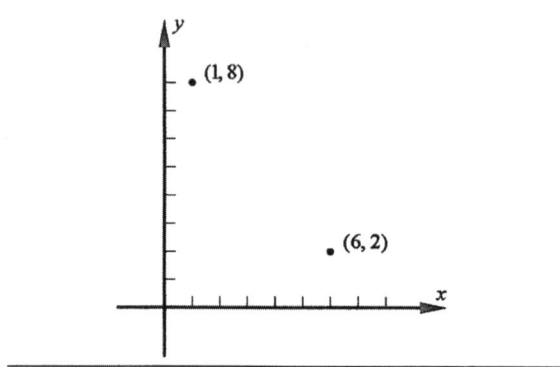

Fig. D.1.7

Derivative Preliminaries 7

First, draw the line determined by our points:

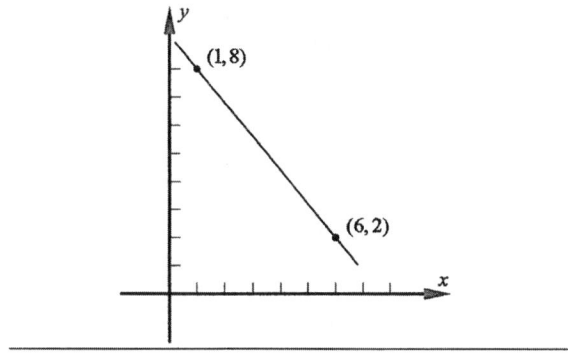

Fig. D.1.8

Then calculate the slope:

$$\text{slope} = \frac{y_2 - y_1}{x_2 - x_1}$$

$$\text{slope} = \frac{2 - 8}{6 - 1}$$

$$\text{slope} = -\frac{6}{5}$$

We need to make two very important points about these two examples. First, notice that in Example 1, the slope of the line was a positive number, while in Example 2, the slope was a negative number.

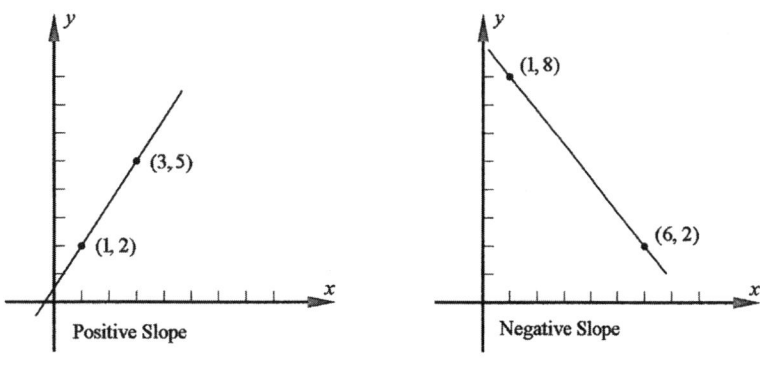

Fig. D.1.9

Second, in neither example did it matter which point we called point 1 or point 2. If we reverse them and again insert the coordinates into the slope equation, we will still get a slope of 3/2 in Example 1 and a slope of -6/5 in Example 2.

We need to consider one more important slope example before we move on to how to interpret the slope once we have found it.

Example 3
Point 1 = (2, 3)
Point 2 = (5, 3)

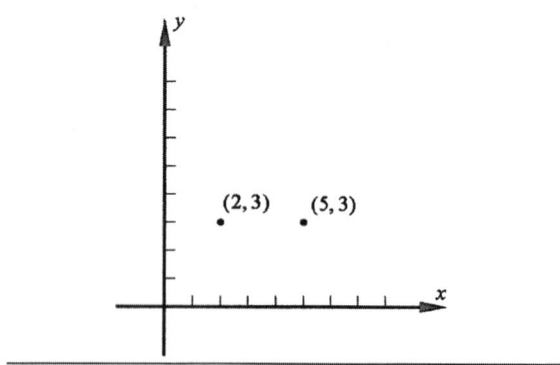

Fig. D.1.10

Again draw the line through the two points:

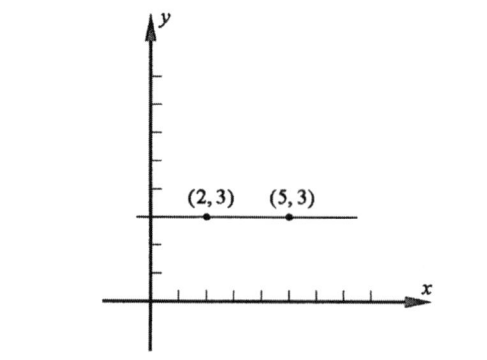

Fig. D.1.11

Notice that because both points have the same y-coordinate, the line drawn is horizontal. Calculating the slope, we find:

$$\text{slope} = \frac{y_2 - y_1}{x_2 - x_1}$$

$$\text{slope} = \frac{3-3}{5-2}$$

$$\text{slope} = \frac{0}{3} = 0$$

Therefore, a horizontal line has zero slope.

▸ Interpreting the Slope of a Line

♦ FYI

Differential calculus is the portion of calculus that involves calculating and using the derivatives of functions.

There are two interpretations of the slope of a line that will be extremely useful as we apply the concept of differential calculus to physical problems. The first of these interpretations is the steepness of a line. In the following graphs, compare the steepness of line #1 to the steepness of line #2:

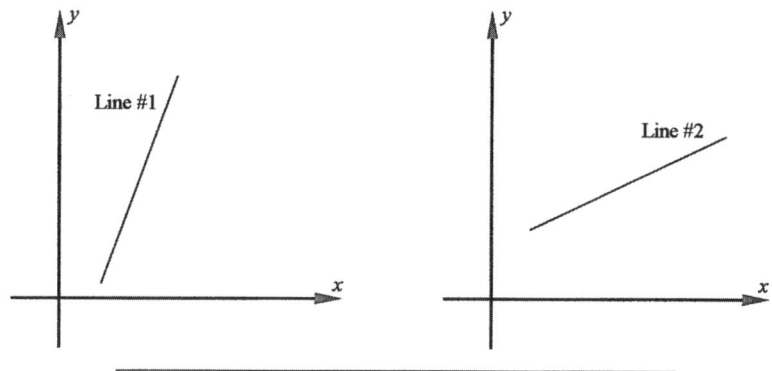

Fig. D.1.12

The second interpretation of the slope is at the heart of a true understanding of differential calculus. Because this interpretation of the slope of a line will be of most use to us in our applications of calculus, we will spend a much larger amount of time analyzing it than we did the first interpretation.

The slope as a rate of change

Let's consider two lines given by two different sets of points:

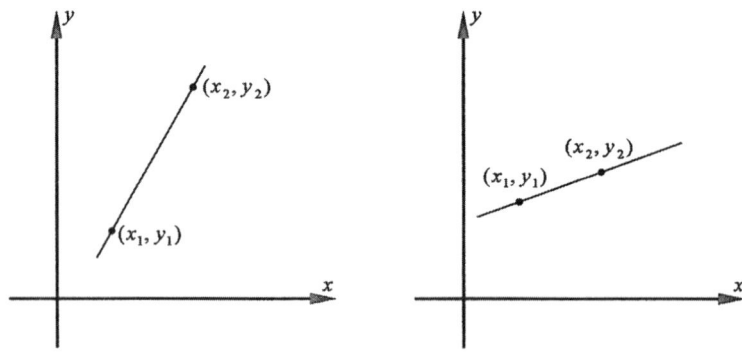

Fig. D.1.13

Remember that x_1 and x_2 are the distances, respectively, of each of our points from $x=0$ along the x-axis.

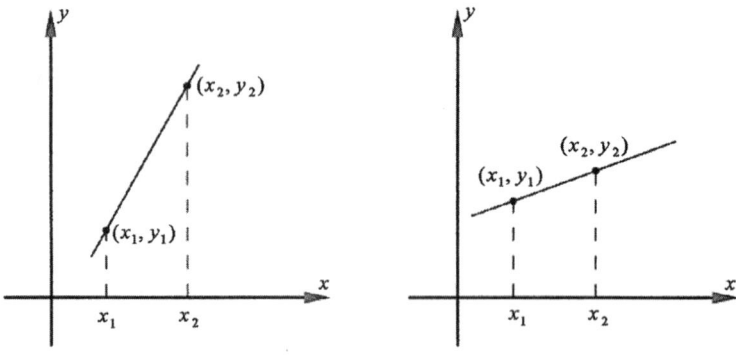

Fig. D.1.14

Notice that in both cases, as we moved from x_1 to x_2 our y-coordinate changed from y_1 to y_2:

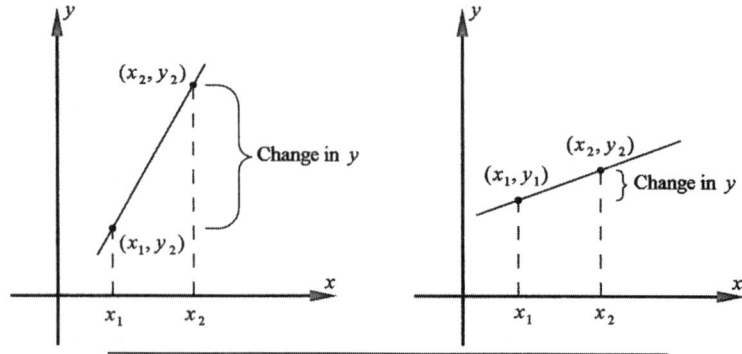

Fig. D.1.15

For line #1, whose slope is larger than line #2, we get a much larger change in the y-coordinate as we move from x_1 to x_2.

> The larger the slope of the line, the larger the change in y as we change x.

In other words, the slope of a line measures how the y-coordinate changes with respect to the x-coordinate.

- A line with a large slope has a large change in y if we change x (as in line #1).
- A line with a small slope has a small change in y if we change x (as in line #2).

Looking again at our slope expression,

$$\text{slope} = \frac{y_2 - y_1}{x_2 - x_1}$$

we can also see this interpretation of the slope as the change in the y-coordinate as we change the x-coordinate. The numerator is the change in the y-coordinate and the denominator is the change in the x-coordinate. In mathematics, it is common to use the capital Greek letter delta, Δ, as a shorthand way of writing "change."

Using this shorthand, we can write our slope expression as

$$\text{slope} = \frac{\Delta y}{\Delta x}$$

We interpret this expression as being a small change in y divided by a small change in x.

Because our goal is to apply these ideas to real-world problems, it is necessary to extend the ideas and interpretations discussed previously to variables other than x and y. We want to be able to find the rate at which a physical quantity, such as displacement, is changing with respect to time. As we will see, our slope interpretations will not change as we extend our technique to include other variables.

▶ Dependent and Independent Variables

Suppose we have an equation such as $y = 2x + 3$. The two variables, x and y, can be given names that distinguish between them. Since we have complete freedom to choose any number to substitute for the variable x, we say that x is the *independent variable*. Notice, however, that while we had complete freedom for the variable x, once a value has been chosen, y is fixed. The variable y is completely dependent on the number substituted for x. Because of this, y is said to be the *dependent variable*.

For example, we may choose x to be 1 and substitute it into the equation

$$y = 2x + 3$$
$$y = 2(1) + 3$$
$$y = 5$$

Or, if we give x a value of 2,

$$y = 2(2) + 3$$
$$y = 7$$

This new way of looking at the relationship between x and y is very useful and extends to other variables as well. In the equation $s = 3t^2 + 4t$, t is the independent variable and s is the dependent variable. The variable s is completely determined once we choose a value for t.

Thinking now of variables and equations in terms of independent and dependent variables, we can extend how we construct graphs and interpret our slopes.

> In graphing, it is conventional to place the independent variable on the horizontal axis and the dependent variable on the vertical axis.

For example, suppose we know that in a certain section of an electric circuit the current i that flows through it changes with time according to

$$i = 3t + 2$$

Notice that the current that is flowing through the section of the circuit depends on the time at which we look at the section. In other words, t is our independent variable and i is our dependent variable. In order to graph the equation, we place t on the horizontal axis and i on the vertical axis:

Derivative Preliminaries   13

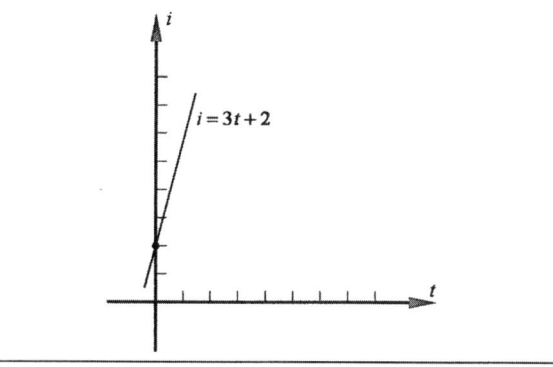

Fig. D.1.16

If we now choose two times to look at the circuit section, $t = 1$ sec and $t = 2$ sec, we can find the current flow at these two times:

$$\text{At } t = 1 \qquad i = 3(1) + 2 = 5$$

$$\text{At } t = 2 \qquad i = 3(2) + 2 = 8$$

> ◆ **FYI**
>
> The unit of amperes, or amps, is normally used for electric currents.

We moved one unit on the horizontal axis (we went from $t = 1$ sec to $t = 2$ sec). As a result of this change, our current went from 5 amps to 8 amps. This means that we have a change of 3 amps of current for every 1 second of time. The current is changing at the rate of 3 amps per second.

If we now take our results and interpret them as the coordinates of two points on our line, we get:

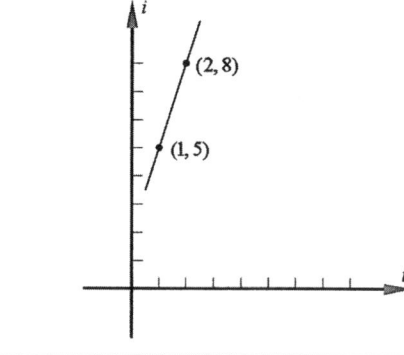

Fig. D.1.17

We can use these coordinates to find the slope of the line:

$$\text{slope} = \frac{i_2 - i_1}{t_2 - t_1}$$

$$\text{slope} = \frac{8\,\text{amps} - 5\,\text{amps}}{2\,\text{sec} - 1\,\text{sec}}$$

$$\text{slope} = \frac{3\,\text{amps}}{1\,\text{sec}}$$

The slope calculation verifies our intuition that the current is changing at the rate of 3 amps per second.

Our next section involves a slight change in notation. It is important to note, however, that our concept of slope will not change. As we will see, this new notation is better suited for defining the concept of a functional derivative.

▶ Functional Notation

There are times when it is actually more convenient and useful *not* to assign the vertical axis to a variable such as y. There are many instances when it is advantageous instead to emphasize that our graph is a function of a certain independent variable. For example, previously we dealt with the equation

$$y = 2x + 3$$

where y was the dependent variable and x was the independent variable. To emphasize the fact that x is the independent variable in the equation, we write the function alternatively as

$$f(x) = 2x + 3$$

The left-hand side of the equation does not mean that f is to be multiplied by x. Rather, it is the notation that we use to express the fact that our expression is a function of the independent variable x.

Functional Notation f(x)

Pronounced "f of x," $f(x)$ is the notation used to indicate that the expression is a function of the independent variable x.

We graph the function as we did before, this time labeling the vertical axis as $f(x)$ instead of y:

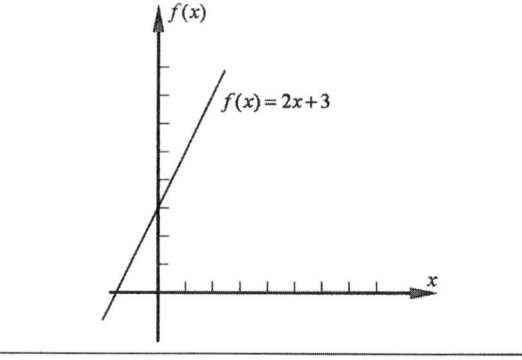

Fig. D.1.18

One of the advantages of this notation is that it explicitly tells us what is being substituted for the independent variable. For example, suppose we have the following function:

$$f(x) = 2x + 1$$

and we want to know the value of the function at $x = 1$. We simply insert 1 into the function everywhere that we see an x:

$$f(x) = 2x + 1$$
$$f(1) = 2(1) + 1$$
$$f(1) = 3$$

We read this as "f of one is equal to three" or "the function evaluated at the point one is equal to three."

Although we could just as easily have substituted the value of 1 into our equation when it was in the form

$$y = 2x + 1$$

the left-hand side of our new notation emphasizes the value of the independent variable that we have substituted, in this case 1. Because this new notation will simplify our derivation of the functional derivative, we need to spend a little more time understanding it.

It is important to note that we can substitute almost anything that we wish into our function, *including letters*. We can evaluate our function at the value p:

$$f(x) = 5x - 7$$
$$f(p) = 5(p) - 7$$
$$f(p) = 5p - 7$$

or at a more elaborate value, such as $p + b$:

$$f(x) = 5x - 7$$
$$f(p+b) = 5(p+b) - 7$$
$$f(p+b) = 5p + 5b - 7$$

During our discussion of functional derivatives, we will need to address situations such as

$$f(x + \Delta x)$$

This is no different from any other substitution that we have discussed so far. If we wanted to evaluate our sample function at $x + \Delta x$, we simply insert this value in the same way that we did 1, or p:

$$f(x) = 5x - 7$$
$$f(x + \Delta x) = 5(x + \Delta x) - 7$$
$$f(x + \Delta x) = 5x + 5\Delta x - 7$$

Another advantage of our new notation is that we can now couple it with our previously discussed concept of slope. Let's return to our graph with two generic points (x_1, y_1) and (x_2, y_2):

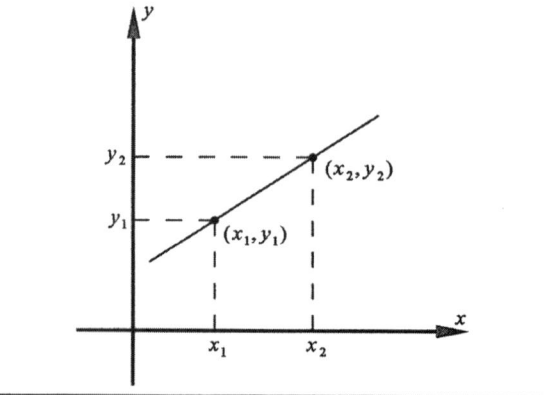

Fig. D.1.19

Because we can generate the vertical coordinate given the horizontal coordinate by using our new notation, we do not have to be given y_1 and y_2. If we know x_1, we can generate the vertical coordinate that is associated with it by substituting x_1 into our function. In other words, we find $f(x_1)$. Likewise, we can find the vertical coordinate associated with x_2 by finding $f(x_2)$. The graph with the two generic points and the line drawn through them now becomes

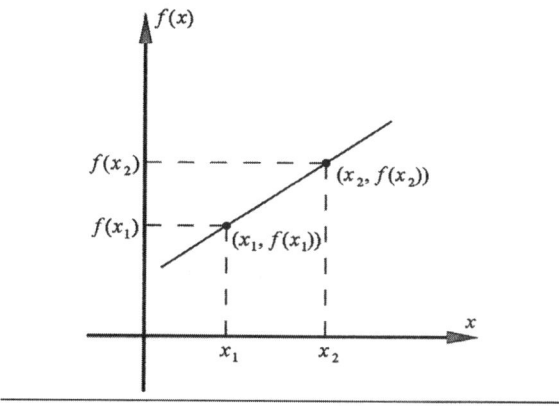

Fig. D.1.20

The slope of the line can still be found by taking the difference in the vertical coordinates and dividing it by the difference in the horizontal coordinates. In this case, however, we will express the vertical coordinates using our new notation:

$$\text{slope} = \frac{f(x_2) - f(x_1)}{x_2 - x_1}$$

Although our notation has changed, the interpretation of the slope as a rate of change remains the same. Because we have expressed the vertical coordinates using our new notation, $f(x)$, we can now interpret our slope as the rate at which the value of the function is changing as we move from x_1 and x_2.

This way of writing the slope of a line, coupled with the concept of the *limit* of a function, discussed in the next section, will yield a method of finding the *derivative* of a function. The derivative of a function is the core of differential calculus.

▶ The Limit of a Function

Suppose we have the function

$$f(x) = 3x + 2$$

and we are interested in the value that it would approach if we let the independent variable, x, get closer and closer to 0. If we choose an arbitrary starting point, we can chart the response of the function as x approaches 0:

$x = 0.1$	$f(0.1) = 3(0.1) + 2 = 2.3$
$x = 0.01$	$f(0.01) = 3(0.01) + 2 = 2.03$
$x = 0.001$	$f(0.001) = 3(0.001) + 2 = 2.003$
$x = 0.0001$	$f(0.0001) = 3(0.0001) + 2 = 2.0003$

We can see from the table that as x approaches 0, the value of the function approaches 2. In mathematical language, we say that the *limit of the function f(x) as x approaches 0 is 2*.

Notice that the result is the same even if we start with values of x less than 0:

$x = -0.1$	$f(-0.1) = 3(-0.1) + 2 = 1.7$
$x = -0.01$	$f(-0.01) = 3(-0.01) + 2 = 1.97$
$x = -0.001$	$f(-0.001) = 3(-0.001) + 2 = 1.997$
$x = -0.0001$	$f(-0.0001) = 3(-0.0001) + 2 = 1.9997$

In other words, provided that our function changes smoothly, the direction from which we approach our value of independent variable does not affect the final value of the function. If we approached $x = 0$ from the right, the function had a value of 2, and if we approached $x = 0$ from the left, the function again had a value of 2.

In equation form, the limit that we found is expressed

$$\lim_{x \to 0} (3x + 2) = 2$$

The question now arises, why bother? From the previous example, it would appear that the limit of a function is no different from the value of the function at a particular value of the independent variable. Why invest the additional amount of work to approach the value of the independent variable from the right, left, etc.? The difference is that the *limit* of the function can usually be found even for values of the independent variable at which the function is undefined.

For example, although we cannot directly substitute a value of $x = 0$ into the function

$$f(x) = \frac{\sin x}{x}$$

since this would involve us dividing by zero, we can let x get as close to zero as we wish. If we charted this approach as we did with the function $3x + 2$, we would see that the function would get closer and closer to 1. In equation form, our limit would take the form

$$\lim_{x \to 0} \frac{\sin x}{x} = 1$$

The theory behind the limits of functions is a very large area of study in and of itself. Because our purpose is to use limits to help us calculate the derivatives of functions, it is sufficient for us to think of a limit as the value of a function as the independent variable approaches a particular value.

Exercises with Solutions

In the following equations, identify the dependent and independent variables:

1. $y = 5x^2 - 7x$

2. $B = 7t^3$

3. $R = 3w + 8$

4. $E = 9s - 11$

Given the following sets of points, calculate the slope of the line that would pass through the points:

5. $(4, 2), (3, 1)$

6. $(2, 5), (8, 1)$

7. $(3, 5), (9, 5)$

8. $(0, 0), (2, 6)$

For each of the following functions, find $f(0)$ and $f(1)$:

9. $f(x) = 2x$

10. $f(x) = 3x^2$

11. $f(x) = 9 - 4x$

12. $f(x) = 4x^2 + 5$

Calculate the limit of the following expressions:

13. $\lim\limits_{x \to 2}(4x+1)$

14. $\lim\limits_{x \to 0}(3x^2 + 4x + 8)$

15. $\lim\limits_{x \to 2} \dfrac{x^2 - 4}{x - 2}$ *Hint:* Recall how to factor the difference of two squares.

Practice Exercises

In the following equations, identify the dependent and independent variables:

1. $y = 4x + 1$

2. $R = 5x^3 + 6$

3. $P = 7t^2 - 9t + 3$

4. $F = t^5$

Given the following sets of points, calculate the slope of the line that would pass through the points:

5. $(6, 1), (5, 0)$

6. $(2, 5), (5, 2)$

7. $(0, 0), (1, 5)$

8. $(1, 3), (5, 3)$

For each of the following functions, find $f(0)$ and $f(1)$:

9. $f(x) = 5x$

10. $f(x) = 4x^3$

11. $f(x) = 7x + 4$

12. $f(x) = \dfrac{2}{x+1}$

Calculate the limit of the following expressions:

13. $\lim\limits_{x \to 1}(5x)$

14. $\lim\limits_{x \to 0}(4x^2 + 3x + 5)$

15. $\lim\limits_{x \to 3} \dfrac{x^3 - 27}{x - 3}$

Differential Calculus

DIFFERENTIAL CALCULUS

2

The Derivative of a Function

 **Connection**

Dependent and independent variables are discussed in Differential Calculus 1.

Although all of the graphs in the previous section had straight lines, in real-world situations it is rare to find two quantities that are related to one another in a linear fashion (i.e., their graph is a straight line). In order to use calculus as a tool in technology, we need to be able to discuss quantities that are related to one another in *non-linear* ways as well as to discuss the rate at which the dependent quantity in the relationship is changing with respect to the independent quantity.

Suppose, instead of a straight line, we have a function such as

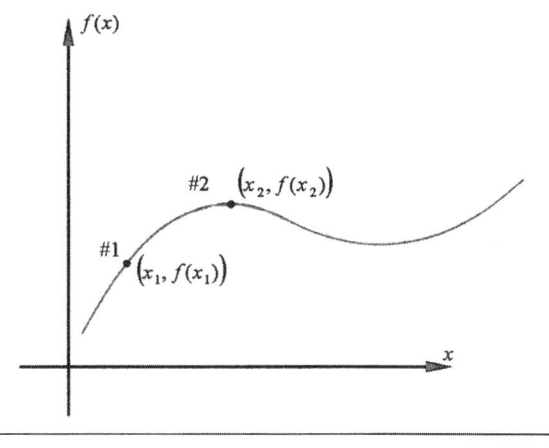

Fig. D.2.1

Notice that we have labeled two arbitrary points on the function using the methods previously discussed. Let's use this sample function to derive an expression that will give us the slope of the tangent line at *any* point on a curve.

If we draw a tangent line at each point, we see that the tangent line at point #1 is steeper than the tangent line at point #2:

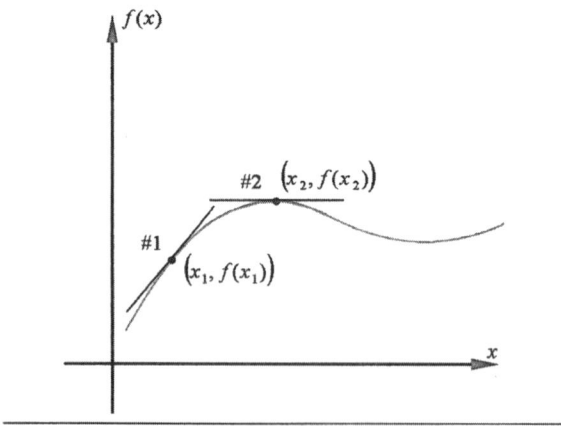

Fig. D.2.2

Recalling that the slope is also the rate of change of the dependent variable with respect to the independent variable, this means that the rate at which the dependent variable is changing with respect to the independent variable is larger at point #1.

We are now at the stage where we can define the *derivative* of the function $f(x)$.

The Derivative of the Function f(x)

When we find the derivative of a function $f(x)$ at a point, we have found

a) the slope of the tangent line drawn at that point
b) the rate of change of the dependent variable with respect to the independent variable at that point

For example, if the curve is a graph of displacement versus time for an object, the derivative at a point would be the velocity of the object at that point.

We have stated what the derivative of a function is, but now how do we actually calculate it?

▶ Finding the Derivative of a Function

Suppose we are interested in the slope of the tangent line (and hence the derivative) at point P:

The Derivative of a Function 25

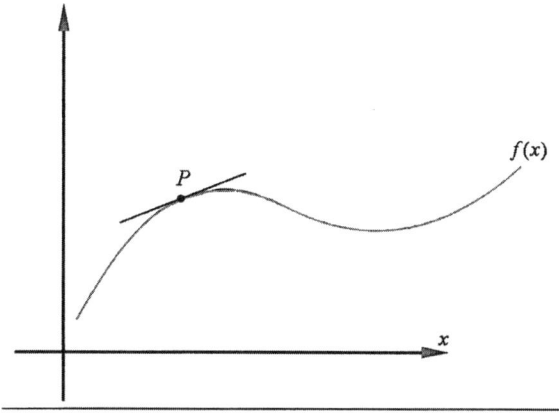

Fig. D.2.3

We start by taking two points, one to the right of P and one to the left,

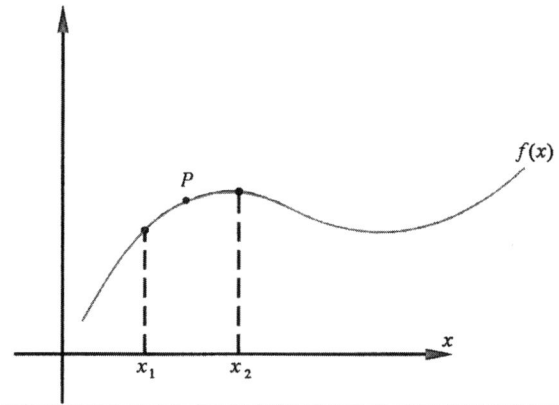

Fig. D.2.4

and drawing the line defined by them:

> ## Connection
>
> Notice that in Fig. D.2.5 we generated the vertical coordinates using the function $f(x)$. For a review of this technique, see page 14.

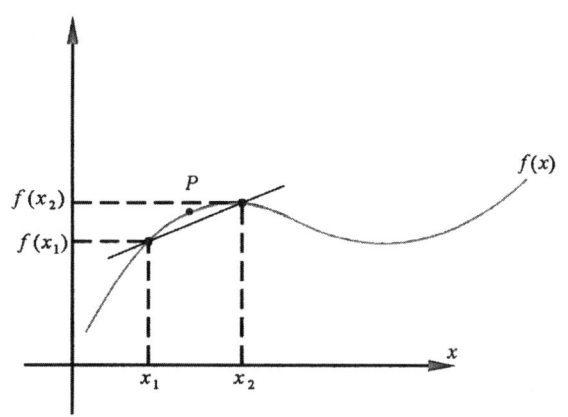

Fig. D.2.5

The slope of this line can be found using our slope expression

$$\text{slope} = \frac{f(x_2) - f(x_1)}{x_2 - x_1}$$

Because we want to use a notation that is as general as possible, let's refer to our first arbitrary point simply as x. If we then express the horizontal distance between our two points as Δx, our second point can be expressed as $x + \Delta x$:

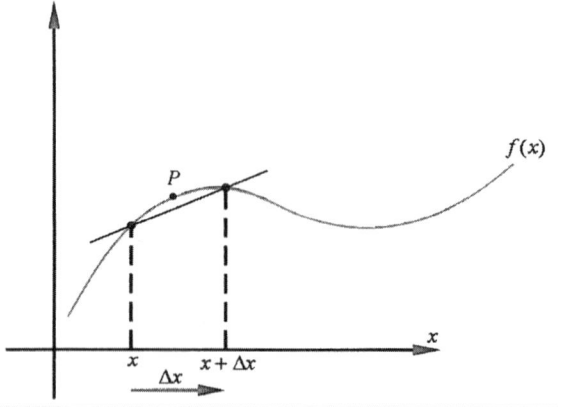

Fig. D.2.6

As before, the value of the function at each point is found by substituting the x value into the function.

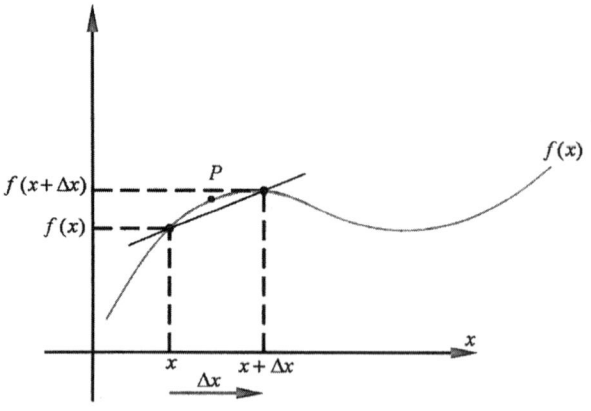

Fig. D.2.7

Thus, the slope expression

$$\text{slope} = \frac{f(x_2) - f(x_1)}{x_2 - x_1}$$

becomes

$$\text{slope} = \frac{f(x+\Delta x) - f(x)}{(x+\Delta x) - x}$$

the denominator of which can be simplified, yielding

$$\text{slope} = \frac{f(x+\Delta x) - f(x)}{\Delta x}$$

The slope of the tangent line determined by the two points gives an approximation to the slope of the line in which we are interested, namely the tangent line at the point P. Obviously, though, our approximation is not a very good one.

We can, however, make the approximation better by choosing two sample points that are closer to the point P:

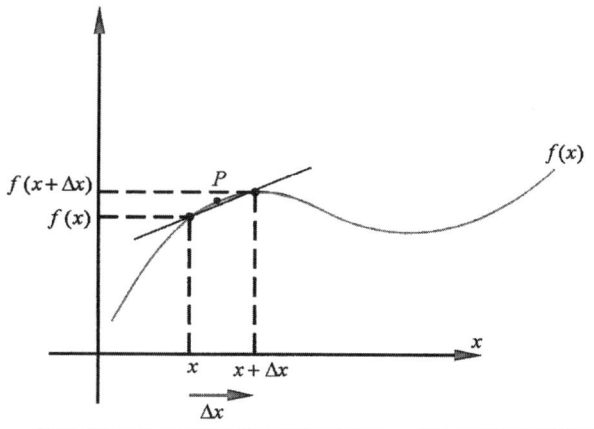

Fig. D.2.8

If we calculated the slope of this new line, it would have a value closer to the slope of the tangent line at P.

We can continue this process of improvement by using sample points that are closer and closer to the point of interest, P:

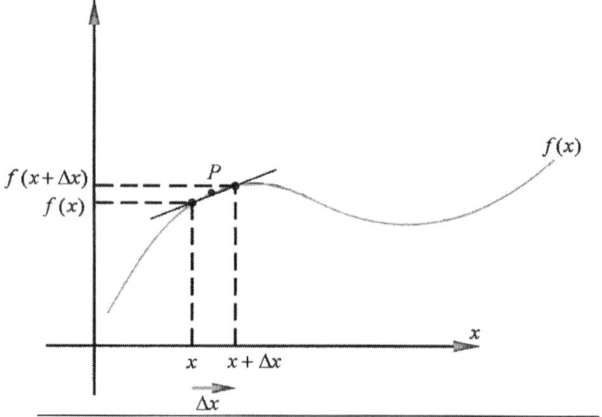

Fig. D.2.9

Thus, the closer we choose the two sample points, the better the approximation to the slope of the tangent line at P. Now let's use the limit concept introduced in the previous section.

Although our two points must remain distinct, it is perfectly permissible for them to get as close to P as we wish. More to the point, it is acceptable for the two sample points to become *infinitely* close to the point P.

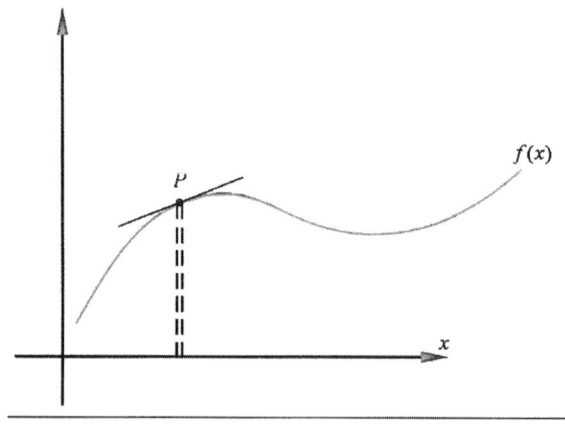

Fig. D.2.10

Because Δx is the distance separating the two sample points, letting them get infinitely close together corresponds to taking the limit as Δx approaches zero:

$$\lim_{\Delta x \to 0}$$

Thus, the slope of the line determined by these two points can be expressed as

$$\text{slope} = \lim_{\Delta x \to 0} \frac{f(x+\Delta x)-f(x)}{\Delta x}$$

Our approximation technique has thus produced the needed result. By finding the slope of the line defined by the two points infinitely close to *P*, we will have found the slope of the tangent line at the point *P*.

Consequently, the slope of the tangent line at *P* can be found using

$$\lim_{\Delta x \to 0} \frac{f(x+\Delta x)-f(x)}{\Delta x}$$

Because the slope of the tangent line at a point on the function is the same thing as the derivative of the function at that point, we have found an expression for the derivative of the function *f(x)*:

The Derivative of the Function f(x)

The derivative of the function *f(x)* is given by

$$\lim_{\Delta x \to 0} \frac{f(x+\Delta x)-f(x)}{\Delta x}$$

It is conventional to use shorthand notations for the derivative of a function. Although there are many different notations, we will only use two in this text. The first of these shorthand notations for the derivative is

$$f'(x)$$

Thus,

$$f'(x) = \lim_{\Delta x \to 0} \frac{f(x+\Delta x)-f(x)}{\Delta x}$$

Having found an equation that will allow us to calculate the derivative of a function, two questions that now arise are

1. For a specific function, how do we use this equation to find the derivative?
2. Once we have found the derivative, what have we actually found?

Example 1

Suppose we have the function $f(x) = x^2$:

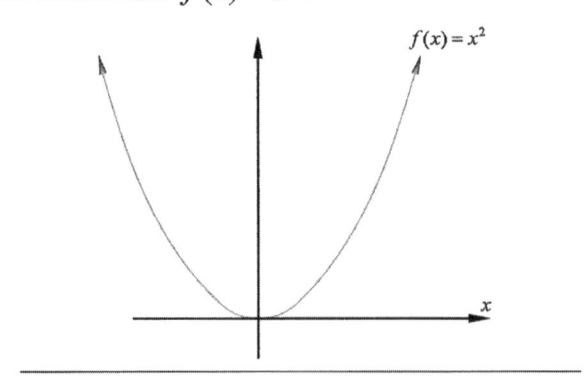

Fig. D.2.11

The definition of the derivative

$$f'(x) = \lim_{\Delta x \to 0} \frac{f(x + \Delta x) - f(x)}{\Delta x}$$

is a recipe that we can use to find the derivative of this function. In order to use it, we must insert the expressions for $f(x + \Delta x)$ and $f(x)$ that correspond to our particular function.

Recall from the section on functional notation that an expression such as $f(x + \Delta x)$ means that we are to take our function,

$$f(x) = x^2$$

and substitute $(x + \Delta x)$ everywhere that we see an x.

Thus,

$$f(x) = x^2$$

becomes

$$f(x + \Delta x) = (x + \Delta x)^2$$

Substituting this result, along with the expression for $f(x)$, the derivative

$$f'(x) = \lim_{\Delta x \to 0} \frac{f(x+\Delta x) - f(x)}{\Delta x}$$

becomes

$$f'(x) = \lim_{\Delta x \to 0} \frac{(x+\Delta x)^2 - x^2}{\Delta x}$$

Because Δx is in the denominator of the right-hand side, we cannot substitute $\Delta x = 0$ directly. We must simplify the right-hand side until Δx is eliminated from the denominator.

First, we expand $(x + \Delta x)^2$,

$$f'(x) = \lim_{\Delta x \to 0} \frac{(x+\Delta x)^2 - x^2}{\Delta x} = \lim_{\Delta x \to 0} \frac{x^2 + 2x\Delta x + (\Delta x)^2 - x^2}{\Delta x}$$

simplifying the numerator,

$$f'(x) = \lim_{\Delta x \to 0} \frac{2x\Delta x + (\Delta x)^2}{\Delta x}$$

factoring Δx out of each term in the numerator,

$$f'(x) = \lim_{\Delta x \to 0} \frac{\Delta x(2x + \Delta x)}{\Delta x}$$

and canceling Δx between numerator and denominator,

$$f'(x) = \lim_{\Delta x \to 0} (2x + \Delta x)$$

With this cancellation, we can now take the limit without having to divide by zero. Finding the limit as Δx approaches zero yields

$$f'(x) = \lim_{\Delta x \to 0} (2x + \Delta x)$$
$$f'(x) = 2x + 0$$
$$f'(x) = 2x$$

The derivative of the function x^2 is $2x$.

Having generated the derivative, we now address the second question as to what we have actually found. When we find a derivative, we find an expression that will yield the slope of the tangent line drawn at any point on the function.

In our example, the derivative of the function

$$f(x) = x^2$$

is

$$f'(x) = 2x$$

Thus, if we are interested in the slope of the tangent line that can be drawn at $x = 2$,

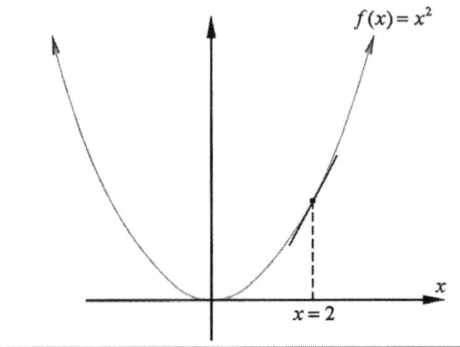

Fig. D.2.12

we simply insert $x = 2$ into the derivative:

$$f'(x) = 2x$$
$$f'(2) = 2(2)$$
$$f'(2) = 4$$

The slope of the tangent line drawn at $x = 2$ is 4:

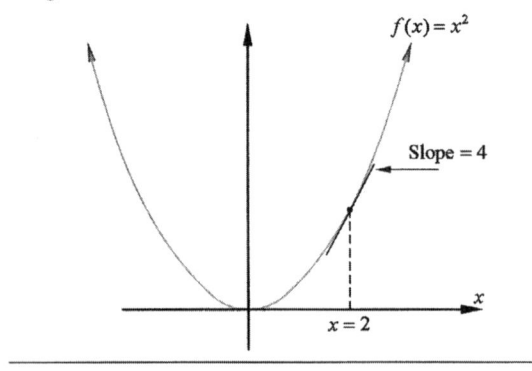

Fig. D.2.13

If we were instead interested in the tangent line drawn at $x = 3$,

$$f'(x) = 2x$$
$$f'(3) = 2(3)$$
$$f'(3) = 6$$

To summarize,

1. We use the definition of the derivative

    $$f'(x) = \lim_{\Delta x \to 0} \frac{f(x + \Delta x) - f(x)}{\Delta x}$$

    as a step-by-step recipe for calculating the derivative of a function. Because Δx is in the denominator of the expression, algebra is normally required before the limit is actually executed.

2. Once we have found the expression for the derivative of the function, we find the slope of the tangent line at any point along the function by substituting that value of x into the derivative expression.

Example 2

Suppose we have the function $f(x) = x^3$:

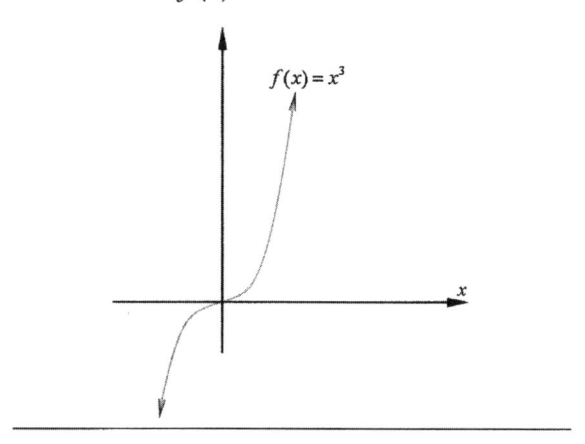

Fig. D.2.14

Remember that to find the derivative, we must always substitute the expression for $(x + \Delta x)$ into the definition. In this case,

$$f(x + \Delta x) = (x + \Delta x)^3$$

34 Differential Calculus

Beginning with the definition and inserting the expressions for $f(x + \Delta x)$ and $f(x)$ for this example,

$$f'(x) = \lim_{\Delta x \to 0} \frac{f(x+\Delta x) - f(x)}{\Delta x}$$

$$f'(x) = \lim_{\Delta x \to 0} \frac{(x+\Delta x)^3 - x^3}{\Delta x}$$

We are again in the position where we cannot directly substitute zero for Δx since this substitution would require us to divide by zero. We need to use algebra to simplify the equation before we actually calculate the limit.

Math Review

Remember that an expression such as

$$(x + \Delta x)^3$$

means to calculate

$$(x+\Delta x)(x+\Delta x)(x+\Delta x)$$

Expanding the cubed term,

$$f'(x) = \lim_{\Delta x \to 0} \frac{(x+\Delta x)^3 - x^3}{\Delta x} = \lim_{\Delta x \to 0} \frac{\left[x^3 + 3x^2\Delta x + 3x(\Delta x)^2 + (\Delta x)^3\right] - x^3}{\Delta x}$$

and simplifying the numerator, we find

$$f'(x) = \lim_{\Delta x \to 0} \frac{3x^2\Delta x + 3x(\Delta x)^2 + (\Delta x)^3}{\Delta x}$$

Factoring Δx off from each term in the numerator yields

$$f'(x) = \lim_{\Delta x \to 0} \frac{\Delta x\left[3x^2 + 3x\Delta x + (\Delta x)^2\right]}{\Delta x}$$

Canceling Δx between numerator and denominator gives us

$$f'(x) = \lim_{\Delta x \to 0}\left[3x^2 + 3x\Delta x + (\Delta x)^2\right]$$

Since Δx is no longer in the denominator, we can take the limit by substituting zero for Δx:

$$f'(x) = 3x^2 + 3x(0) + (0)^2$$
$$f'(x) = 3x^2$$

The derivative of the function x^3 is $3x^2$.

If we would now like to know the slope of the tangent line that can be drawn at, for example, $x = 1$, we substitute 1 into the derivative everywhere that we see an x:

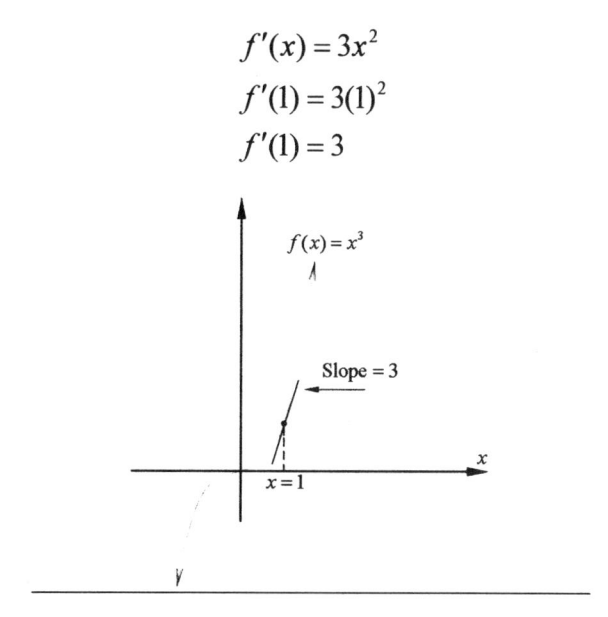

$$f'(x) = 3x^2$$
$$f'(1) = 3(1)^2$$
$$f'(1) = 3$$

Fig. D.2.15

Exercises with Solutions

Use the definition of the derivative

$$f'(x) = \lim_{\Delta x \to 0} \frac{f(x + \Delta x) - f(x)}{\Delta x}$$

to calculate the derivatives of the following functions:

1. $f(x) = 2x$

2. $f(x) = 3x^2$

3. $f(x) = 5x - 3$

4. $f(x) = 4x^3$

5. $f(x) = 6x^2 + 7x$

Practice Exercises

Use the definition of the derivative

$$f'(x) = \lim_{\Delta x \to 0} \frac{f(x+\Delta x) - f(x)}{\Delta x}$$

to calculate the derivatives of the following functions:

1. $f(x) = 2x + 7$

2. $f(x) = x^2 + 6$

3. $f(x) = x^3 + 2x$

4. $f(x) = \dfrac{3}{x}$

5. $f(x) = 6x^3 + 4x^2$

6. $f(x) = \sqrt{x}$

7. $f(x) = \dfrac{7x^2}{4}$

8. $f(x) = x^4$

9. $f(x) = 6 - 3x$

10. $f(x) = \dfrac{6}{\sqrt{x}}$

DIFFERENTIAL CALCULUS

3

Rules for Finding Derivatives

From the two examples in Differential Calculus 2 - *The Derivative of a Function*, we see that calculating derivatives starting from the definition is both difficult and time consuming. We were forced into several steps of simplification and algebra in order to reach our goal. If we were forced into that many steps for a function as simple as

$$f(x) = x^2$$

you can imagine how difficult the algebra becomes if we deal with a function such as

$$f(x) = (x^3 + 5x^2 + 3)(4x^2 - 7x + 11)$$

Because our eventual goal is to be able to apply the concept of a derivative to real-world situations such as projectile motion and circuit analysis, we need to find a much simpler way of calculating derivatives than using the Δx method. This section shows much easier and faster ways of finding the derivatives of functions. As we will see, the shortcut that we use to find the derivative of the function quickly will depend upon the type of function. For example, simple polynomials will have a different shortcut from the one used for trigonometric functions. Because we do not know what sorts of functions may arise in our physical problems, we will cover a range of different functions that often arise in the process of solving science and engineering problems.

Note: To keep the derivatives in this chapter as "user-friendly" as possible, we will use words to describe our derivative rules throughout this section. In addition, a formula for the derivative of each type of function is placed in a box under the definition box. A table summarizing the various derivative formulae given throughout the chapter can be found at the end of this chapter.

Rule 1 – Simple Polynomials

Finding the Derivatives of Simple Polynomials

1. Term by term, take the exponent on the term, bring it down, and make it a coefficient in front of the term.
2. Lower the original exponent by one power.
3. The derivative of any constant is zero.

Note: In Step 1, if there is already a numerical coefficient in front of the term, it should be multiplied by our new coefficient.

Simple Polynomials:
$$f(x) = ax^n + bx^{n-1} + ...$$
Derivative:
$$f'(x) = nax^{n-1} + (n-1)bx^{n-2} + ...$$

Let's take this rule and apply it to the function

$$f(x) = 3x^2 + 4$$

After we apply the rule, we will check our answer by finding the derivative using the Δx method.

Example 1

Using the rule for simple polynomials, find the derivative of

$$f(x) = 3x^2 + 4$$

Solution First, we address the $3x^2$ term. Our rule says that we should take the exponent, in this case a 2, and bring it down in front of the term. Since there is already a 3 in front of the term, we will multiply the 2 and the 3 together. Next, we lower the original exponent, 2, by one power. Writing out our steps, the derivative of the $3x^2$ portion of our function becomes

$$2(3)x^{2-1} = 6x$$

Now addressing the constant term of 4, the third step of our rule tells us that the derivative of any constant term is zero. As we will see when we check our result using the Δx method, what is really underlying this part of the rule is that we will have a 4 and a -4 that will cancel one another, leaving a result of zero for this portion of the derivative.

Putting the two pieces together, we get

$$f(x) = 3x^2 + 4$$
$$f'(x) = 6x + 0$$
$$f'(x) = 6x$$

Let's now recalculate the derivative of the function using the Δx method.

Beginning with the definition

$$f'(x) = \lim_{\Delta x \to 0} \frac{f(x + \Delta x) - f(x)}{\Delta x}$$

we get

$$f'(x) = \lim_{\Delta x \to 0} \frac{3(x + \Delta x)^2 + 4 - (3x^2 + 4)}{\Delta x}$$

> **Math Review**
>
> Remember that the expression $f(x + \Delta x)$ means that we are to substitute $x + \Delta x$ into our function everywhere that we see an x.

Expanding the squared term and removing the parentheses on the second term, we find

$$f'(x) = \lim_{\Delta x \to 0} \frac{3\left[x^2 + 2x\Delta x + (\Delta x)^2\right] + 4 - 3x^2 - 4}{\Delta x}$$

Distributing the 3 in front of the parentheses yields

$$f'(x) = \lim_{\Delta x \to 0} \frac{3x^2 + 6x\Delta x + 3(\Delta x)^2 + 4 - 3x^2 - 4}{\Delta x}$$

We can now cancel terms in the numerator of the expression. Notice that the constant term of 4 cancels at this stage, as we mentioned previously.

After canceling terms, we find

$$f'(x) = \lim_{\Delta x \to 0} \frac{6x\Delta x + 3(\Delta x)^2}{\Delta x}$$

Factoring Δx off of each term in the numerator gives us

$$f'(x) = \lim_{\Delta x \to 0} \frac{\Delta x(6x + 3x\Delta x)}{\Delta x}$$

Canceling Δx between numerator and denominator then yields

$$f'(x) = \lim_{\Delta x \to 0} (6x + 3\Delta x)$$

Now that we have executed the algebra necessary to remove Δx from the denominator, we can safely find the limit as Δx approaches zero:

$$f'(x) = \lim_{\Delta x \to 0} (6x + 3\Delta x)$$
$$f'(x) = 6x + 3(0)$$
$$f'(x) = 6x$$

Comparing our result with the derivative found using our rule for polynomials, we see that the rule does indeed produce the same derivative as the Δx method.

Let's now take our rule and find the derivatives of a few simple polynomials.

Example 2
Find the derivative of the function

$$f(x) = 5x^3 + 4x^2 + 7x + 3$$

We will find the derivative of the function term by term. For each of the terms in the function, we will bring down the exponent in front and then lower the exponent on the term by one power. If there is already a constant in front of the term, we will multiply it by our exponent.

For the $5x^3$ term,

1. bring the 3 down in front of the term
2. multiply the 3 by the 5 which is already there, getting 15
3. lower the exponent of 3 down to 2

The derivative of $5x^3$ is $15x^2$.

For the $4x^2$ term,

1. bring the 2 down in front of the term
2. multiply the 2 and the 4 together, getting 8
3. lower the exponent of 2 to 1

The derivative of $4x^2$ is $8x$.

For the $7x$ term, we need to point out a couple of things while using our rule. Remember that there is really an exponent of 1 on the x. Conventionally, we do not usually write in the numeral. During the process of applying our rule, we always take the exponent and lower it by one power. When we take the exponent of 1 that is on the x and lower it by one power, this will drop the exponent to zero.

To find the derivative of $7x$,

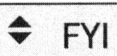

1. bring the exponent of 1 that is on the x down in front of the term
2. multiply it by the 7 that is already there, getting 7
3. lower the exponent of 1 by one power, bringing it to zero

Since $x^0 = 1$, we get 7 as the derivative.

Example 3

Find the derivative of

$$f(x) = 6x^4 - 8x^3 + 5x^2 - 2.75$$

Solution Using our rule for polynomials:

$$f'(x) = 24x^3 - 24x^2 + 10x$$

There are two points for this particular example that are worthwhile to note. First, notice that the second term in the function is *negative* $8x^3$. Because the term is negative, we multiplied the exponent of 3 that we brought down by negative 8. Second, even though 2.75 is not an integer, it still cancelled in the process of calculating the derivative since it is a constant.

Example 4

If the tangent line to the function

$$f(x) = 3x^2 + x + 5$$

is drawn at $x = 1$, find the slope of this tangent line.

Math Review

A number raised to the zero power is equal to one. For example,

$5^0 = 1$
$7^0 = 1$

FYI

Another way to think about the derivatives of expressions such as $7x$, $8x$, etc., is that we simply retain the coefficient on the term.

Math Review

Remember that the x term in a function such as

$f(x) = 3x^2 + x + 5$

has an understood coefficient of 1.

42 Differential Calculus

Solution Remember that when we find the derivative of a function, we have found an expression that will give us the slope of the tangent line to the curve at any point in which we are interested. For this function, the derivative is

$$f'(x) = 6x + 1$$

Since we are interested in the slope of the tangent line drawn at $x = 1$, all we have to do is substitute a 1 everywhere that we see an x in the derivative:

$$f'(x) = 6x + 1$$
$$f'(1) = 6(1) + 1$$
$$f'(1) = 7$$

The slope of the tangent line drawn at $x = 1$ is 7.

Example 5

In this last example before we move to the next derivative rule, we will have the opportunity to review some algebraic properties.

Find the derivative of the function

$$f(x) = \sqrt[3]{x^2} + \frac{5}{x^2}$$

Solution Although at first glance it does not appear that we will be able to use our rule on this function, we will indeed be able to use it. However, we must first rewrite the function using two algebraic properties. We will address each of the terms individually in order to illustrate these properties.

First, let's look at the radical term. We can rewrite this term using fractional exponents. To do this, we must take the power on the x, in this case a 2, and make it the numerator of our fractional exponent. We then take the root that we are calculating, in this case a third root, and make it the denominator of our fractional exponent. Rewriting our radical expression in terms of fractional exponents, we get

$$\sqrt[3]{x^2} = x^{\frac{2}{3}}$$

Math Review

$$\sqrt[m]{x^n} = x^{\frac{n}{m}}$$

Notice that because the variable x is now written to a power, we will be able to use the rule for calculating its derivative.

Now addressing the fractional term in our function, remember that we can always move terms between the numerator and the denominator of a fraction by

Math Review

$$\frac{1}{x^n} = x^{-n}$$

changing the sign on the exponent. Since we have raised the x in the denominator of the fraction to the second power, we can move the x into the numerator of the expression by changing the 2 to *negative* 2:

$$\frac{5}{x^2} = 5x^{-2}$$

Again, because our variable is raised to a power, we can apply our rule.

Combining the radical and the fractional terms, we can rewrite our original function as

$$f(x) = \sqrt[3]{x^2} + \frac{5}{x^2}$$

$$f(x) = x^{\frac{2}{3}} + 5x^{-2}$$

Our rule can now be used even though the exponents on each term are not positive whole numbers. We still follow our strategy of bringing down the exponent and lowering the original exponent by one power:

$$f(x) = x^{\frac{2}{3}} + 5x^{-2}$$

$$f'(x) = \frac{2}{3}x^{\frac{2}{3}-1} + 5(-2)x^{-2-1}$$

$$f'(x) = \frac{2}{3}x^{-\frac{1}{3}} - 10x^{-3}$$

Exercises with Solutions

Find the derivatives of the following functions using the rule for polynomials:

Section 1

1. $f(x) = 6x^2 + 4x + 3$

2. $f(x) = 5x - 4$

3. $f(x) = 7x^5 - 2x^4$

44 Differential Calculus

4. $f(x) = -8x^2 + 15$

5. $f(x) = 2x - 5x^3$

6. $f(x) = -4x + 19$

7. $f(x) = 14x^2 + 7x - 6.75$

8. $f(x) = 8.3x + 7$

9. $f(x) = 9$

10. $f(x) = 12x - 4.3x^2 + 5x^4$

Section 2

Note: The exercises in this section are slightly more difficult and may require some algebra.

11. $f(x) = \dfrac{3}{x^2}$

12. $f(x) = \sqrt[3]{x^4}$

13. $f(x) = \sqrt{x}$

14. $f(x) = x^3 + \sqrt{x}$

15. $f(x) = \dfrac{4}{\sqrt[3]{x}}$

16. $f(x) = 4x^3 - 5x^2 + \dfrac{7}{x} + 3$

17. $f(x) = \dfrac{x^2 - 4}{2}$

18. $f(x) = \dfrac{4}{x^5} + \dfrac{3}{x^2} + 5$

19. $f(x) = \dfrac{3x^2 + 4x}{3}$

20. $f(x) = \dfrac{x^3}{\sqrt{x}}$

Practice Exercises

Find the derivatives of the following functions using the rule for polynomials:

1. $f(x) = 6x^2 + 7x + 5$

2. $p(x) = 6(x^2 + 24x)$

3. $f(x) = 8x^3 + 4x^2 + 10x$

4. $h(x) = \dfrac{x^2 - 4}{2}$

5. $f(x) = 14x + 5$

6. $d(x) = x^3 - 64$

7. $f(x) = \sqrt{x}$

8. $f(x) = \dfrac{5}{2}x^2$

9. $f(x) = \dfrac{6}{x^2}$

10. $f(x) = \pi$

11. $p(x) = 8x^5 - 4x^2$

12. $f(x) = e^2$

13. $b(x) = 40x^4 - 8x$

14. $f(x) = 3.5x^2 + 7.2x$

15. $f(x) = \dfrac{5}{\sqrt{x}}$

16. $p(x) = \dfrac{3.2}{\sqrt{x}}$

17. $f(x) = \sqrt{x} + \dfrac{4}{\sqrt{x}}$

18. $f(x) = \dfrac{x^5 + 6x^4 + 3x^3 + 7x^2 - 5x}{6}$

19. $g(x) = 6x^3 + 4.75$

20. $f(x) = \dfrac{x\sqrt{x}}{\sqrt[3]{x}}$

Rule 2 – Functions Involving the Sine and Cosine

As we move on to our second rule, it is again important to bear in mind that we can still use the Δx method to find the derivatives of these functions. These rules will simply produce the derivative more quickly.

Finding the Derivative of the Sine Function

1. Take the argument of the sine function (whatever is inside the parentheses) and make it the argument of the cosine function. *Do not change what is inside the parentheses in any way.*
2. Multiply in front of the cosine function by the derivative of whatever is inside the parentheses.

Sine functions:
$$f(x) = \sin[g(x)]$$
Derivative:
$$f'(x) = g'(x)\cos[g(x)]$$

Math Review

In the function
$$f(x) = \sin(3x)$$
$3x$ is the *argument* of the sine function.

Example 6
Find the derivative of
$$f(x) = \sin(3x^2 + 4x)$$

Solution
$$f'(x) = (6x+4)\cos(3x^2 + 4x)$$

Notice that the $(3x^2 + 4x)$ that was the argument of the sine function is now the argument of the cosine function. Also, the term in front of the cosine function, $6x + 4$, is the derivative of $(3x^2 + 4x)$ (Step 2 in our rule).

Example 7
Find the derivative of
$$f(x) = 4\sin(6x^3 + 5x^2 - 3x)$$

Solution Since our function has 4 as a coefficient, we will have 4 multiplying our derivative. Applying our rule for the sine function,
$$f'(x) = 4(18x^2 + 10x - 3)\cos(6x^3 + 5x^2 - 3x)$$

Again, the argument of the sine function became the argument of the cosine function without changing it at all. Next, we multiplied in front of the cosine by the derivative of this argument.

As we move on to the cosine function, note that the rule is almost identical to the one used to find the derivative of the sine function.

Finding the Derivative of the Cosine Function

1. Take the argument of the cosine function (whatever is inside the parentheses) and make it the argument of the sine function. *Do not change what is inside the parentheses in any way.*
2. Multiply in front of the sine function by the derivative of what is inside the parentheses.
3. As a last step, multiply your answer by negative 1.

Cosine functions:
$$f(x) = \cos[g(x)]$$
Derivative:
$$f'(x) = -g'(x)\sin[g(x)]$$

Example 8
Find the derivative of

$$f(x) = \cos(5x^2 + 6x)$$

Solution Applying our rule for the cosine function yields:

$$f'(x) = -(10x + 6)\sin(5x^2 + 6x)$$

There are several items that should be noted here. First, notice that the $(5x^2 + 6x)$ that was the argument of the cosine function is now the argument of the sine function. Second, the $(10x + 6)$ that multiplies the sine function is the derivative of $(5x^2 + 6x)$ (Step 2 in our rule). Third, we have multiplied our final answer by negative 1 (Step 3 in our rule).

Example 9
Find the derivative of
$$f(x) = -6\cos(4x^3 - 8x^2)$$

Solution Since there is a factor of -6 multiplying the function, we also need to have a factor of -6 in our derivative. Applying our rule, we get

$$f'(x) = -(-6)(12x^2 - 16x)\sin(4x^3 - 8x^2)$$
$$f'(x) = 6(12x^2 - 16x)\sin(4x^3 - 8x^2)$$

Notice that the negative sign on the 6 and the negative sign inserted due to Step 3 in our rule cancel one another, leaving the sign in front of the derivative positive in this case.

▶ Rule 3 – Functions Involving the Exponential Function

This section deals with functions that contain e, the base of the natural logarithms. For example,

$$f(x) = e^{2x+1}$$

Finding the Derivatives of Exponential Functions

1. Recopy the exponential function *without changing the function in any way*.
2. Multiply in front of the function by the derivative of the exponent to which e has been raised.

Exponential functions:
$$f(x) = e^{g(x)}$$
Derivative:
$$f'(x) = g'(x)e^{g(x)}$$

Example 10
Find the derivative of
$$f(x) = e^{2x+1}$$

Solution Following our rule, we first recopy the function. After this, we multiply in front by the derivative of $(2x + 1)$, which is 2:

$$f'(x) = 2e^{2x+1}$$

Just as in the case of our previous functions, by calculating the derivative of this particular exponential function, we have found an expression that will give us the slope of the tangent line that can be drawn to the function. If we would like the slope of the tangent line that can be drawn at $x = 1$, we simply substitute the number 1 everywhere that we see an x in our derivative:

$$f'(x) = 2e^{2x+1}$$
$$f'(1) = 2e^{2(1)+1}$$
$$f'(1) = 2e^3$$
$$f'(1) = 40.17$$

> **Math Review**
>
> Remember that e is an actual number ($e \cong 2.71828$) that can be raised to a power just as we do any other number.

Example 11

Find the derivative of the function

$$f(x) = 8e^{6x-10}$$

Solution Notice that because there is a factor of 8 multiplying the function, we will have a factor of 8 multiplying our derivative. Applying our rule for exponential functions gives

$$f'(x) = 6\left(8e^{6x-10}\right)$$
$$f'(x) = 48e^{6x-10}$$

▶ Rule 4 – Functions Involving the Natural Logarithm

In this section, we address functions that are related to the exponential functions discussed in Rule 3. In that section, we learned how to find the derivatives of functions that involved e, the base of the natural logarithm. In this section, our focus is on the rule for finding the derivatives of functions that contain the natural logarithm itself, such as

$$f(x) = \ln(4x + 1)$$

Finding the Derivative of the Natural Logarithm

1. Take the argument of the logarithm and place it under 1.
2. Multiply the fraction from Step 1 by the derivative of the argument.

Functions Containing the Natural Logarithm:
$$f(x) = \ln[g(x)]$$

Derivative:
$$f'(x) = \frac{g'(x)}{g(x)}$$

Example 12
Find the derivative of the function

$$f(x) = \ln(4x+1)$$

Solution First, we take the argument of the logarithm, $4x+1$, and place it under 1,

$$\frac{1}{4x+1}$$

according to Step 1 in our rule.

Next, we multiply this fraction by the derivative of the argument $4x+1$ to yield the derivative of the function:

$$f'(x) = 4 \cdot \frac{1}{4x+1}$$
$$f'(x) = \frac{4}{4x+1}$$

Example 13
Find the derivative of the function

$$f(x) = 6\ln(5x^2 + 3x)$$

Solution We place the argument of the logarithm, $5x^2 + 3x$, in the denominator,

$$\frac{1}{5x^2 + 3x}$$

and then multiply by the derivative of the argument, yielding

$$(10x+3) \cdot \frac{1}{5x^2 + 3x} = \frac{10x+3}{5x^2 + 3x}$$

Since there is a factor of 6 in the original function,

$$f(x) = 6\ln(5x^2 + 3x)$$

the derivative also contains a factor of 6:

$$f'(x) = 6 \cdot \frac{10x+3}{5x^2 + 3x}$$

$$f'(x) = \frac{60x + 18}{5x^2 + 3x}$$

Exercises with Solutions

Find the derivative of each of the following functions:

Section 1

1. $f(x) = \sin(3x)$
2. $f(x) = \ln(2x)$
3. $f(x) = \sin x$
4. $f(x) = 6\ln(7x^2 + 9x)$
5. $f(x) = \cos(3x^4)$
6. $f(x) = -\ln x$
7. $f(x) = \cos x$
8. $f(x) = -e^{3x}$
9. $f(x) = e^{7x+8}$
10. $f(x) = \cos(7x^3 - 5x^2 + 10x)$
11. $f(x) = e^x$
12. $f(x) = 8\sin(x^2)$

Section 2 - Mixed Types

The exercises in this section are slightly more difficult and may require a little algebra before trying to find the derivative.

13. $f(x) = \sqrt{x}$

14. $f(x) = \dfrac{8}{x^5} - 4x^3$

15. $f(x) = 6\cos(9x^3 - 5x^2)$

16. $f(x) = e^{7x^3 - 5x}$

17. $f(x) = e^{\sqrt{x}}$

18. $f(x) = \cos(5x - 3) + \sqrt[3]{x^4}$

19. $f(x) = \dfrac{5}{x^6} + \sin(9x)$

20. $f(x) = \cos(5x^2 - 12x) + \sin(6x - 1)$

21. $f(x) = e^{7x - 3x^4} + \sin(8x) + 2x^5$

22. $f(x) = \ln(2x) + \sin x$

23. $f(x) = \cos(2x^2) - \ln(6x - 1)$

24. $f(x) = 5e^{4.1x}$

25. $f(x) = 3.2\ln(2x + 1)$

Practice Exercises

Find the derivative of each of the following functions:

1. $f(x) = \sin(2x)$
2. $f(x) = e^{6x}$
3. $f(x) = \sin(6x^2 + 5x)$
4. $f(x) = e^{7x^2 + 5x}$
5. $f(x) = \sin(4 - x^2)$
6. $f(x) = e^{-4x^2}$
7. $f(x) = \sin\left(\dfrac{4}{x^2}\right)$
8. $f(x) = e^{\sqrt{x}}$
9. $f(x) = \sin(\sqrt{x})$
10. $f(x) = \ln(9x^2 + 3x)$
11. $f(x) = \cos(5x)$
12. $f(x) = \sin(5x - 6) + e^{8x}$
13. $f(x) = \cos(9x^2 + 4x)$
14. $f(x) = \cos(9x^2 + x + 3) + \sin(6x^3 + 14x)$
15. $f(x) = \ln(8x)$
16. $f(x) = e^{15x^2 + 7x} - \sin(6x^2 - 2x + 10)$
17. $f(x) = \cos\left(\dfrac{4}{x^5}\right)$
18. $f(x) = e^{6x+5} + 7\ln(x^3)$
19. $f(x) = \cos(\sqrt[4]{x})$
20. $f(x) = \sin(7x^3 + 4.5x) - \cos(16x + 5)$

Rules for Combinations of Functions

In this section, we discuss rules for combinations of the simple functions covered in Rules 1, 2, 3, and 4. As we cover these combination rules, it is again important to bear in mind that the formal definition of the derivative could still be used on these combinations.

▸ Rule 5 – The Product Rule

From basic algebra, we know that the term *product* implies multiplication. We use the *Product Rule* whenever we have two expressions multiplied together, such as

$$f(x) = (3x^2 + 4x + 8)(7x^3 + 5x^2)$$

Math Review

If we wanted to simplify the derivative to the right, we would use the method of common factors.

The Product Rule

We can find the derivative of a function that is a product of two expressions by taking the first expression times the derivative of the second expression, plus the second expression times the derivative of the first expression.

The Product of Two Functions:
$$f(x) = g(x)h(x)$$

The Product Rule:
$$f'(x) = g(x)h'(x) + h(x)g'(x)$$

Example 14

Find the derivative of

$$f(x) = (3x^2 + 4x + 8)(7x^3 + 5x^2)$$

Solution In this case, our first term is $3x^2 + 4x + 8$ and the second term is $7x^3 + 5x^2$. The Product Rule is a step-by-step recipe for finding the derivative of this function. We are instructed to write down the first term and then multiply it by the derivative of the second term. Next, we are to add on the second term multiplied by the derivative of the first term. Thus,

$$f'(x) = (3x^2 + 4x + 8)(21x^2 + 10x) + (7x^3 + 5x^2)(6x + 4)$$

At this point, we could simplify our derivative. However, it is instructional to leave it in its current form so we can see each term in the Product Rule.

Example 15
Find the derivative of

$$f(x) = (5x^4 - 3x^2)(8x^6 + 4x^5 + 2x^3)$$

Solution In this example, our first term is $5x^4 - 3x^2$ and our second term is $8x^6 + 4x^5 + 2x^3$. Applying the Product Rule, we get

$$f'(x) = (5x^4 - 3x^2)(48x^5 + 20x^4 + 6x^2) + (8x^6 + 4x^5 + 2x^3)(20x^3 - 6x)$$

Notice that we have again left the derivative unsimplified so we can see each term of the Product Rule.

We can use the Product Rule whenever we have two terms multiplied together, not just if the terms are simple polynomials. In this next example, one of the terms is not a simple polynomial. However, we will still apply the Product Rule in the same manner.

Example 16
Find the derivative of

$$f(x) = e^{5x+3}(2x^3 + 4x)$$

Solution In this case, our first term is e^{5x+3} and our second term is $2x^3 + 4x$. Applying the Product Rule yields

$$f'(x) = e^{5x+3}(6x^2 + 4) + (2x^3 + 4x)(5e^{5x+3})$$

Now that we have learned the rule for two expressions multiplied together, let's discuss the rule for two expressions divided by one another.

▶ Rule 6 – The Quotient Rule

Like the Product Rule, the Quotient Rule is a step-by-step recipe that allows us to find the derivative of a function such as

$$f(x) = \frac{4x^3 + 3x^2 - 7x}{5x^2 + 4x}$$

Math Review

In a fraction such as
$$\frac{a}{b}$$
a is the numerator of the fraction and b is the denominator of the fraction.

The Quotient Rule

To find the derivative of a function that contains two expressions divided by one another, take the denominator times the derivative of the numerator, *minus* the numerator times the derivative of the denominator, *all* over the denominator squared.

The Quotient of Two Functions:
$$f(x) = \frac{g(x)}{h(x)}$$

The Quotient Rule:
$$f'(x) = \frac{h(x)g'(x) - g(x)h'(x)}{[h(x)]^2}$$

Example 17
Find the derivative of the function

$$f(x) = \frac{4x^3 + 3x^2 - 7x}{5x^2 + 4x}$$

Solution In this case, the numerator is $4x^3 + 3x^2 - 7x$ and the denominator is $5x^2 + 4x$. Applying the Quotient Rule gives us

$$f'(x) = \frac{(5x^2 + 4x)(12x^2 + 6x - 7) - (4x^3 + 3x^2 - 7x)(10x + 4)}{(5x^2 + 4x)^2}$$

We have again left the derivative in an unsimplified form so that each term in the Quotient Rule is apparent.

Example 18
Like its counterpart the Product Rule, we can use the Quotient Rule on quotients that contain expressions that are not simple polynomials, such as

$$f(x) = \frac{\sin(2x)}{\cos(3x)}$$

Inserting the numerator, sin(2x), and the denominator, cos(3x), into the Quotient Rule yields

$$f'(x) = \frac{(\cos 3x)(2\cos 2x) - (\sin 2x)(3)(-\sin 3x)}{(\cos 3x)^2}$$

If we wished, we could now use various relationships between the sine and cosine functions to simplify our derivative.

▸ Using the Quotient Rule to Find the Derivatives of the Tangent, Secant, Cosecant, and Cotangent Functions

Previously in this chapter, we discussed the rules for calculating the derivatives of functions containing the sine and cosine functions. In this section, we can draw upon the Quotient Rule to develop the rules for finding the remaining four basic trigonometric functions: tangent, cotangent, secant, and cosecant.

Since each of these trigonometric functions can be expressed as fractions that contain the sine and cosine function,

$$f(x) = \tan x = \frac{\sin x}{\cos x}$$

$$f(x) = \cot x = \frac{\cos x}{\sin x}$$

$$f(x) = \sec x = \frac{1}{\cos x}$$

$$f(x) = \csc x = \frac{1}{\sin x}$$

we can use the rules for the sine and cosine coupled with the Quotient Rule to find the derivatives of the four remaining trigonometric functions.

The Derivative of the Tangent Function

We begin by expressing the tangent function in terms of the sine and cosine:

$$f(x) = \tan x$$

$$f(x) = \frac{\sin x}{\cos x}$$

The derivative of the tangent function can now be found using the Quotient Rule:

$$f(x) = \frac{\sin x}{\cos x}$$

$$f'(x) = \frac{(\cos x)(\cos x) - (\sin x)(-\sin x)}{(\cos x)^2}$$

58 Differential Calculus

The derivative can be simplified yielding a simple result:

$$f'(x) = \frac{(\cos x)(\cos x) - (\sin x)(-\sin x)}{(\cos)^2}$$

$$f'(x) = \frac{\cos^2 x + \sin^2 x}{\cos^2 x}$$

$$f'(x) = \frac{1}{\cos^2 x} = \sec^2 x$$

Thus,

$$\boxed{\frac{d}{dx}(\tan x) = \sec^2 x}$$

The Derivative of the Secant Function

By expressing the secant function as a fraction involving the cosine function,

$$f(x) = \sec x$$

$$f(x) = \frac{1}{\cos x}$$

we can find the derivative using the Quotient Rule:

$$f(x) = \frac{1}{\cos x}$$

$$f'(x) = \frac{(\cos x)(0) - (1)(-\sin x)}{(\cos x)^2}$$

Simplifying,

$$f'(x) = \frac{(\cos x)(0) - (1)(-\sin x)}{(\cos x)^2}$$

$$f'(x) = \frac{\sin x}{\cos^2 x} = \frac{1}{\cos x} \cdot \frac{\sin x}{\cos x}$$

$$f'(x) = \sec x \tan x$$

Thus,

$$\boxed{\frac{d}{dx}(\sec x) = \sec x \tan x}$$

The Derivative of the Cotangent Function

Expressing the cotangent function as a fraction and using the Quotient Rule, we find

$$f(x) = \cot x = \frac{\cos x}{\sin x}$$

$$f'(x) = \frac{(\sin x)(-\sin x) - (\cos x)(\cos x)}{(\sin x)^2}$$

$$f'(x) = \frac{-\sin^2 x - \cos^2 x}{\sin^2 x} = \frac{-(\sin^2 x + \cos^2 x)}{\sin^2 x}$$

$$f'(x) = \frac{-1}{\sin^2 x} = -\csc^2 x$$

Thus,

$$\boxed{\frac{d}{dx}(\cot x) = -\csc^2 x}$$

The Derivative of the Cosecant Function

Applying the Quotient Rule to the cosecant function gives us

$$f(x) = \csc x = \frac{1}{\sin x}$$

$$f'(x) = \frac{(\sin x)(0) - (1)(\cos x)}{(\sin x)^2}$$

$$f'(x) = \frac{-\cos x}{\sin^2 x} = -\frac{\cos x}{\sin x} \cdot \frac{1}{\sin x}$$

$$f'(x) = -\cot x \csc x$$

Thus,

$$\boxed{\frac{d}{dx}(\csc x) = -\cot x \csc x}$$

▶ Rule 7 – The Power Rule

Our last type of combination function in this section involves functions in which an expression in brackets has been raised to a power, such as

$$f(x) = (3x^2 + 5x)^5$$

The Power Rule

1. Take the exponent that is on the term in brackets, bring it down, and make it a coefficient, lowering the original exponent by one.
2. Multiply this new expression by the derivative of what is inside the brackets.

A Function Containing a Bracket Raised to a Power:
$$f(x) = [g(x)]^n$$

The Power Rule:
$$f'(x) = n[g(x)]^{n-1}[g'(x)]$$

Example 19
Find the derivative of

$$f(x) = (3x^2 + 5x)^5$$

Solution First, we bring the exponent of 5 down and place it as a coefficient in front of the parentheses and lower the exponent down to 4:

$$5(3x^2 + 5x)^4$$

Second, we multiply this expression by the derivative of $3x^2 + 5x$. Coupling these two pieces together, our derivative take the form

$$f'(x) = 5(3x^2 + 5x)^4 (6x + 5)$$

Example 20
Find the derivative of

$$f(x) = (7x^3 - 6x^2 + 3x)^8$$

Solution Applying the Power Rule yields

$$f'(x) = 8(7x^3 - 6x^2 + 3x)^7 (21x^2 - 12x + 3)$$

Derivative Rules

1. Simple polynomials: $f(x) = ax^n + bx^{n-1} + ...$

    Derivative: $f'(x) = nax^{n-1} + (n-1)bx^{n-2} + ...$

2. Sine functions: $f(x) = \sin[g(x)]$

    Derivative: $f'(x) = g'(x)\cos[g(x)]$

3. Cosine functions: $f(x) = \cos[g(x)]$

    Derivative: $f'(x) = -g'(x)\sin[g(x)]$

4. Exponential functions: $f(x) = e^{g(x)}$

    Derivative: $f'(x) = g'(x)e^{g(x)}$

5. Functions containing the natural logarithm: $f(x) = \ln[g(x)]$

    Derivative: $f'(x) = \dfrac{g'(x)}{g(x)}$

6. The product of two functions: $f(x) = g(x)h(x)$

    The Product Rule: $f'(x) = g(x)h'(x) + h(x)g'(x)$

7. The quotient of two functions: $f(x) = \dfrac{g(x)}{h(x)}$

    The Quotient Rule: $f'(x) = \dfrac{h(x)g'(x) - g(x)h'(x)}{[h(x)]^2}$

8. A function containing a bracket raised to a power: $f(x) = [g(x)]^n$

    The Power Rule: $f'(x) = n[g(x)]^{n-1}[g'(x)]$

Exercises with Solutions

Section 1

Use the Product Rule, Quotient Rule, or Power Rule to find the derivatives of these functions involving simple polynomials:

1. $f(x) = (3x^2 + 8x)(5x^3 + 7x^2 - 6x)$

2. $f(x) = (4x^5 + 5x^2 + 3x)(9x^2 - 6x)$

3. $f(x) = \dfrac{x^3 + 4x^2}{7x^4 + 3x^3}$

4. $f(x) = \dfrac{9x - 1}{x^5 - 6x^3}$

5. $f(x) = (4x^3 + 5x^2 - 6x)^4$

6. $f(x) = (7x^2 - 11x + 10)^6$

7. $f(x) = (6x^3 + 5x)(4x^4 - 7x^2)$

8. $f(x) = (5x^7 - 6x^4 + 9x^2)^5$

9. $f(x) = \dfrac{5x^4 + 3x^2 - 9x}{4x^2 - 7.5x + 3.2}$

10. $f(x) = (5x^3 + 7x)^4 + \dfrac{3x^3 - 4x^2}{9x^2 - 8x + 2}$

Section 2

Use the Product Rule, Quotient Rule, or Power Rule to find the derivatives of the following functions.

Note: The Product Rule, Quotient Rule, and Power Rule do not change if we use a choice of independent variable other than x.

11. $f(x) = \dfrac{e^{4x^3+6x^2}}{3x^4}$

12. $f(t) = (7t^2 - 6t + 9)\sin(5t^3 + 8t)$

13. $f(t) = (5t + e^{7t})^4$

14. $f(x) = \dfrac{\sin(3x-2)}{\sqrt{x^5}}$

15. $f(p) = (6p + \sqrt{p})(5p^3 + 8p^2 - 12p)$

Practice Exercises

Section 1

Use the Product Rule, Quotient Rule, or Power Rule to find the derivatives of these functions involving simple polynomials:

1. $f(x) = (6x^2 + 3x)(9x^3 + 4x^2)$

2. $f(x) = \dfrac{x^2 + \sqrt{x}}{x^3 - 6x^2 + 5x}$

3. $f(x) = (15x^2 + 8x - 3)(4x^3 - 10x^2)$

4. $f(x) = \dfrac{8 - 6x}{5x + 7}$

5. $f(x) = (6x + \sqrt{x})(11x^3 - 7x^2)$

6. $f(x) = \dfrac{5.2x^2 + 6}{8x^3 + 7x^2}$

7. $f(x) = (6 - 9x^2)(4x^7 + 8x^4 + 2x)$

8. $f(x) = (6x^2 + 5x)^3$

9. $f(x) = (10x^2 + 9x + 3)(40x^3 + 16x^2 + 2)$ 10. $f(x) = (7x^3 - 9x^2 + 10x)^6$

11. $f(x) = \dfrac{9x^2 + 5x}{7x^3 - 4x^2}$

12. $f(x) = (10x^4 + \sqrt[3]{x})^5$

13. $f(x) = \dfrac{6x^3 + 10x^2 - 11x}{15x^2 - 8x + 7}$

14. $f(x) = \left(8x^3 + \dfrac{8}{x^2}\right)^7$

Section 2

Use the Product Rule, Quotient Rule, or Power Rule to find the derivatives of the following functions.

Note: The rules do not change if we use a choice of independent variable other than x.

15. $g(x) = e^{9x} \sin(3x^2 + 5x)$

16. $g(x) = \dfrac{x^2 + 6x}{e^{5x}} + (9x^2 - 7x + 3)^5$

17. $f(t) = \dfrac{\cos(15t + 3)}{\sin(4t^2 + 8t)}$

18. $f(t) = \sqrt{t^3 + 5t^2 + e^{7t}}$

19. $f(x) = (x^2 + e^{6x})^4$

20. $p(r) = (9r^2 + 5r - 3)^4 (7r^3 + 10r + 17)^6$

Summary Exercises with Solutions

The purpose of this set of exercises is to bring together all of the rules for finding derivatives that we have discussed in the preceding sections.

Find the derivative of each of the following functions:

1. $f(x) = 7x^3 - 5x$

2. $f(t) = (3t - 5)^7$

3. $f(t) = \sin(9t^3 - 4t)$

4. $f(x) = \dfrac{3x^4 - 7x^2}{4x - 3}$

5. $f(b) = e^{3b^2 - 5b + 2}$

6. $f(x) = \sqrt{2x + 5}$

7. $f(x) = \dfrac{5}{x^4} + \cos(4x)$

8. $f(t) = (6t^2 - 4t)(5t^3 + 7t - 3)$

9. $f(d) = 8d^4 - 6d^3 + 3d$

10. $f(x) = \dfrac{e^{3x}}{2x - 7}$

11. $f(x) = 9\ln(x^2)$

12. $f(t) = e^{5t}(4t^2 - 11t)$

13. $f(w) = \sin(5w) + \dfrac{4w - 12}{w^2 + 2w}$

14. $f(x) = \dfrac{\cos(8x + 2)}{\sqrt{x}}$

15. $f(x) = \sqrt[3]{9x^2 + 5x - 6}$

DIFFERENTIAL CALCULUS

4

Differentials I – An Alternative Method of Expressing Derivatives

In this chapter, we will discuss an alternative way of expressing our derivatives. As we will see, this new notation is preferable when applying the concepts of differential calculus to most physical situations. As we begin this chapter, it is important to note that we will not change the methods by which we calculate our derivatives, only the notation that we use to express the answer.

Suppose we have the following generic function with a typical tangent line drawn at a point on the function:

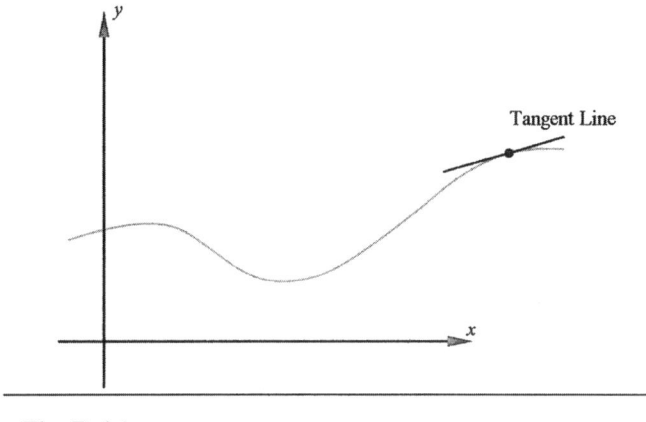

Fig. D.4.1

Previously, we discussed the fact that we can express the slope of a line as

$$\text{slope} = \frac{\Delta y}{\Delta x}$$

and we interpreted it as meaning the rate at which the dependent variable, y, was changing with respect to the independent variable, x. Also, we discussed the fact that the derivative of a function evaluated at a particular point is the same as the slope of the tangent line at that point. Our purpose here is to take this interpretation of the slope of a line and express it using the new notation of *differentials*. As we will see, this new notation is particularly well suited to physical applications of the derivative.

▶ Differentials

Differentials are infinitely small amounts of change in a certain quantity. Since the Greek letter delta (Δ) is used to represent change, we use the letter d, the first letter of delta, to represent an infinitely small amount of change.

For example, Δx is how we express the amount of distance that we move along the x-axis. If we wanted to express an *infinitely* small amount of movement along the x-axis, we would write dx.

If we wanted to express the amount that the velocity of an object changed, we would write Δv. If the object's velocity changed by an infinitely small amount, we would write dv.

We can now apply the concept of a differential to the slope of our tangent line. Remember that the slope of a tangent line is the rate at which y is changing, at our point of interest, with respect to x. In other words, this slope is the rate at which y is changing as we move a small amount along the x-axis.

Consequently, if we move an *infinitely* small amount of distance along the x-axis, an infinitely small amount of change along the y-axis will result. Where we write

$$\frac{\Delta y}{\Delta x}$$

to represent a *small* change in y resulting from a change in x, we write

$$\frac{dy}{dx}$$

to represent an *infinitesimal* change in y resulting from an infinitesimal change in x.

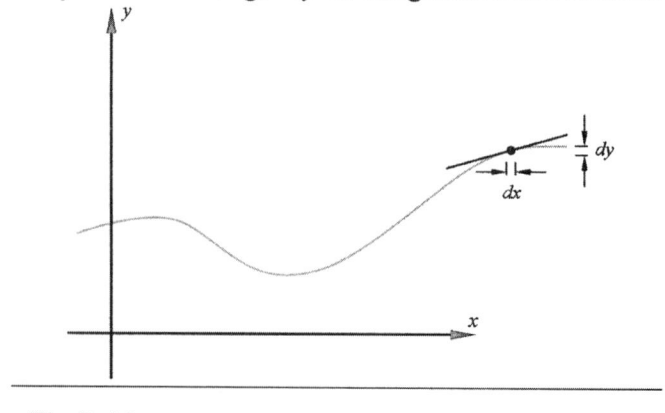

Fig. D.4.2

This infinitesimal change in y due to an infinitesimal change in x corresponds to the discussion in Differential Calculus 1 in which we found the slope of the tangent line at a point using two infinitely close points. The dy term corresponds to the vertical change between the two points, and the dx term corresponds to the horizontal change between the two points. In that chapter, we showed that using two infinitesimally separated points, we can actually find the slope of the tangent line at any point on a curve and that this slope was the same thing as the value of the derivative of the function at that point.

Thus,

$$\frac{dy}{dx}$$

is just an alternative method of expressing the derivative of a function. When the function is expressed as $f(x)$ we express the derivative as $f'(x)$. When the function is expressed as y, we use $\frac{dy}{dx}$ to express the derivative.

For example, we know that we can express the following function as either

$$f(x) = 3x^2 + 4x \qquad (1)$$

or

$$y = 3x^2 + 4x \qquad (2)$$

We find the derivatives of Equations (1) and (2) in the same way. If we find the derivative of the function when it is in the form of Equation (1), we express the derivative as

$$f'(x) = 6x + 4$$

If we find the derivative of the function when it is in the form of Equation (2), we express the derivative as

$$\frac{dy}{dx} = 6x + 4$$

This alternative way of writing derivatives is advantageous for two reasons. First, it emphasizes the relationship between derivatives and slopes. Second, it explicitly states that the dependent variable is changing with respect to the independent variable. This second point is extremely beneficial to us as we apply the concepts of differential calculus to physical problems.

Example 1
Calculate the derivative of the function

$$y = 7x^3 + 5x$$

Solution   Using the rule for simple polynomials and expressing our answer using our differential notation, our derivative takes the form

$$\frac{dy}{dx} = 21x^2 + 5$$

In the next example, let's start with a function that has a different dependent and independent variable. As we apply the concepts of differential calculus to physical problems, it is important for us to realize that we may have different variables than simply x and y.

Example 2
Find the derivative of

$$B = e^{6t+4}$$

Solution   In this case, B is our dependent variable and t is our independent variable. Applying the exponential rule and expressing the derivative using the notation from this section, we get

$$\frac{dB}{dt} = 6e^{6t+4}$$

Example 3
Find the derivative of

$$y = 3\ln x$$

Solution   Recalling the rule for the derivative of the natural logarithm, we find

$$\frac{dy}{dx} = 3 \cdot \frac{1}{x} = \frac{3}{x}$$

▸ Conclusion

There are times when our function will be written using functional notation, $f(x)$, and times when it will be written using y. The manner in which we express the derivative changes, but not the rules that we use to find the derivatives. When we find the derivative of a function, regardless of the notation used, we are finding the slope of the tangent line to the curve and the rate at which the dependent variable is changing with respect to the independent variable.

Exercises with Solutions

Find the derivatives of the following functions. Express each derivative using differential notation.

1. $y = 6x$

2. $y = 7x^2 + 3x + 2$

3. $y = e^{4x}$

4. $B = 5t^4 + 7t^2$

5. $E = \sin(5x)$

6. $x = 5\cos(2t)$

7. $q = \cos t$

8. $p = (4t^2 + 3t)^5$

9. $v = \dfrac{5}{t^3}$

10. $y = \sqrt{x}$

11. $y = \ln x$

12. $B = \tan t$

13. $y = e^{9x^2} + \ln(x^2 + 7x)$

14. $B = (x^3 + 7x)\ln(6x)$

15. $y = \sec x$

Practice Exercises

Find the derivatives of the following functions. Express each derivative using differential notation.

1. $B = 6t^2 + 5t$

2. $R = (b^3 + 7b^2 + 2b)(6b^5 + 9b^4)$

3. $F = g^3 - g^2 + 5g + 6$

4. $G = \dfrac{8}{\sqrt[3]{c^2}}$

5. $P = \sqrt{x}$

6. $y = x^2 + \sin(7x^4 + 8x^3) + e^{9x}$

7. $E = \cos(9t^2)$

8. $M = \sqrt[3]{n^2} + e^{\sqrt{n}}$

9. $P = (v^3 + 7v^2)^5$

10. $V = \dfrac{\sin(8t)}{\cos(6t)}$

11. $y = e^{9x^2} + 5x$

12. $B = \ln(x^3)$

13. $x = 3t + 4.9t^2$

14. $Q = \sqrt{8t^2 - 11t}$

15. $c = \sin(4w^2 + 6w)$

16. $S = \dfrac{y^5 + 4y^2}{\sin(3y)}$

17. $k = \dfrac{y^2 + 7y + 3}{4y^2 + 8y}$

18. $y = e^{15x^2 + 6x} + \tan x$

19. $j = \dfrac{5}{p^6} + \sec p$

20. $x = \dfrac{r^2}{\sin(5r)} + e^{6r}$

DIFFERENTIAL CALCULUS

5

Higher-Order Derivatives

In this chapter, we continue our discussion of derivatives by finding higher-order derivatives. As we will see, finding higher-order derivatives simply involves us finding the derivative of a function more than once.

▶ Higher-Order Derivatives Using Functional Notation

Example 1
Suppose we have the function

$$f(x) = 3x^4 + 2x^3$$

Finding the derivative of this function yields

$$f'(x) = 12x^3 + 6x^2$$

The derivative that we have found is referred to as a *first derivative* since we have found the derivative of the function one time. If we now take the derivative of this first derivative, the result is referred to as a *second derivative*:

$$f'(x) = 12x^3 + 6x^2$$
$$f''(x) = 36x^2 + 12x$$

Notice that we attached two prime symbols to our second derivative, indicating that we have found the derivative of the same function twice. Following the same pattern, it is possible for us to find the *third derivative* of the function by taking the derivative of the second derivative:

$$f''(x) = 36x^2 + 12x$$
$$f'''(x) = 72x + 12$$

▶ Higher-Order Derivatives Using Differentials

We can also express higher-order derivatives using the differential notation discussed in Differential Calculus 4 - *Differentials I*.

Example 2

Suppose we have the function

$$y = 2x^5 - 7x^2 + 9$$

Expressed using differential notation, our first derivative takes the form

$$\frac{dy}{dx} = 10x^4 - 14x$$

If we now find the second derivative of this function, the method of finding the derivative does not change, but it is expressed using the notation

$$\frac{d^2y}{dx^2} = 40x^3 - 14$$

It is important to note that the 2's that appear in the second derivative do not imply that something is being squared. Rather, this is simply the notation that we use to alert the reader that a second derivative has been found.

If we now find the third derivative of the function using this notation, it takes the form

$$\frac{d^3y}{dx^3}$$

We can assign an interpretation to the higher-order derivatives of this section by remembering that derivatives are rates of change. When we find the first derivative of a function, we find the rate at which the dependent variable is changing with respect to the independent variable. Continuing this interpretation, when we find the *second* derivative of a function, we are finding the rate of change of the first derivative. Since an important interpretation of the derivative is as a slope, an interpretation of the second derivative is as the rate at which the slope of the tangent line is changing at a point on the function. Similarly, the third derivative is the rate of change of the second derivative, and so on.

Higher-Order Derivatives

Exercises with Solutions

Section 1

Find the first and second derivatives of the following functions using the notation from Example 1:

1. $f(x) = 5x^6 - 7x^3 + 9x$

2. $f(x) = 3x^4 + 6x^3$

3. $f(x) = 5x^3 - 9$

4. $f(x) = \sqrt{x}$

5. $f(x) = \dfrac{9}{x^2}$

Section 2

Find the first and second derivatives of the following functions using the notation from Example 2:

6. $y = 5x^3 + 6x^2$

7. $y = 3x^{-4}$

8. $y = \dfrac{3}{4}x^5 - \sqrt{x}$

9. $y = e^{6x}$

10. $y = \sin(2x)$

78 Differential Calculus

Section 3

Find the first and second derivatives of the following functions. Express each derivative using the appropriate dependent and independent variables: dB/dt, dP/dr, etc.

11. $x = 5t^4 - 8t^3$

12. $F = 7s^3 + 8s - 3$

13. $P = 4w^6 - \dfrac{2}{3}w^4$

14. $B = \sqrt{t}$

15. $Z = \dfrac{4}{g^3} + 7g$

Practice Exercises

Find the first and second derivatives of the following functions using the appropriate notation from Examples 1 and 2:

1. $f(x) = 9x^3 + 5x^2 + 6x$

2. $y = e^{7x^2}$

3. $y = 5x^4 + 7x^3$

4. $f(x) = x^6 + \sin(5x) + \sqrt{x}$

5. $B = 10t^5 - 6t^4 + 7$

6. $g(x) = \dfrac{5}{\sqrt[3]{x}}$

7. $F = \sin(6g)$

8. $B = \dfrac{4f^2 - 169}{2f - 13}$

9. $f(x) = e^{5x}$

10. $f(x) = 2x^{4.5}$

11. $g(x) = 4x + 11$

12. $f(x) = \sqrt{x+3}$

13. $y = \dfrac{x^5 + 3x^2}{4}$

14. $y = \dfrac{x^2}{\sqrt{x}}$

15. $K = \cos(5r - 2)$

16. $H = e^{8t} + t^{2.3}$

17. $f(x) = (6x - 3)^8$

18. $f(x) = 10$

19. $h(x) = \dfrac{2x + 3}{x - 1}$

20. $f(x) = \dfrac{x^3 + e^{10x}}{5}$

DIFFERENTIAL CALCULUS

6

Implicit Differentiation

In this section, we will discuss a method of finding derivatives of functions whose form is not immediately apparent. Because this technique is based on the Power Rule, we open this section with a review of this derivative rule.

Example 1
Find the derivative of the function

$$y = (3x^2 + 6x + 5)^4$$

Solution Let's review the two steps required to calculate this type of derivative.

> **Math Review**
>
> A *coefficient* is a constant that multiplies a function.
>
> In the function
> $$y = 7x^2$$
> 7 is the coefficient.

First, we take the exponent on the parentheses, bring it down and make it a coefficient in front, and lower the original exponent by one power.

Second, we multiply this new expression by the derivative of the expression inside the parentheses.

$$\frac{dy}{dx} = 4(3x^2 + 6x + 5)^3 (6x + 6)$$

As we learn how to *implicitly differentiate*, we will be executing these same two steps. The only difference is that the form of the function will not be completely apparent to us. In other words, the dependent variable will not be isolated in the function whose derivative we would like to find.

▶ Implicit Differentiation

Suppose that instead of having a function in which the dependent variable is isolated, such as Example 1, we have a function such as

$$y^3 + 6x^2 = 5x^3 \qquad (1)$$

In this case, we do not have y expressly written in terms of x. Although it is possible to rearrange Equation (1) and solve for y, suppose now that we have a more elaborate function, such as

$$y^3 x^4 + e^{6y} = \cos(x^2)\sin(y^5)$$

Even if it is possible, the amount of time required to isolate y in such an expression would be enormous. *Implicit differentiation* gives us a method of finding the derivatives of such expressions without going through the intense algebra required to isolate y.

Example 2
Implicitly differentiate the expression

$$y^5 + 6x^4 = 3x^2 \qquad (2)$$

Solution Our strategy is to begin on the left side of the expression and find the derivative of each term individually. However, although the terms $6x^4$ and $3x^2$ can be handled relatively easily, the y^5 term requires more attention.

Remembering that y is a function of x, we realize that the term y^5 represents a function of x that has been raised to the fifth power. Consequently, to find the derivative of such a term, we apply the Power Rule.

Beginning with the Power Rule, the first step is to take the exponent, 5, make it a coefficient, and lower the exponent to 4:

$$5y^4$$

The second step in the application of the Power Rule is to multiply by the derivative of the term that was raised to the fifth power. Unfortunately, we do not know what y is in terms of x. However, we do know how to write an expression for the derivative of y with respect to x, namely

$$\frac{dy}{dx}$$

Thus, the derivative of y^2 would be $2y\frac{dy}{dx}$, the derivative of y^3 would be $3y^2\frac{dy}{dx}$, etc. Putting it together, the implicit derivative of y^5 is

$$5y^4 \frac{dy}{dx}$$

If we now take this result and couple it with the other required derivatives of Equation (2) involving x, we get

$$5y^4 \frac{dy}{dx} + 24x^3 = 6x \qquad (3)$$

We can now use Equation (3) to find the derivative of our unknown function y. Since the derivative of the function y is dy/dx, we will have our derivative if we execute the required algebra on Equation (3) to isolate the dy/dx term:

$$5y^4 \frac{dy}{dx} + 24x^3 = 6x$$

$$5y^4 \frac{dy}{dx} = 6x - 24x^3$$

$$\frac{dy}{dx} = \frac{6x - 24x^3}{5y^4}$$

Using the methods of implicit differentiation, we were able to find the derivative of the function y without knowing its exact form at the beginning of the problem.

To summarize our method:

Implicit Differentiation

1. Beginning on the left side of the expression, find the derivative of each term. Remember also to find the derivatives of those terms on the right side of the equal sign.
2. Anytime you take the derivative of y raised to a power, calculate the derivative using the normal methods for finding derivatives but remember to attach a dy/dx to the result.
3. Execute the algebra necessary to isolate dy/dx.

Let's now use the method of implicit differentiation on a few examples.

Example 3

Implicitly differentiate

$$y^3 + 6x^2 = e^{7x}$$

Solution First, we begin on the left side and calculate the derivative of each term independently, attaching *dy/dx* whenever we calculate the derivative of a *y*, yielding

$$3y^2 \frac{dy}{dx} + 12x = 7e^{7x}$$

Executing the algebra required to isolate *dy/dx* yields

$$3y^2 \frac{dy}{dx} + 12x = 7e^{7x}$$

$$3y^2 \frac{dy}{dx} = 7e^{7x} - 12x$$

$$\frac{dy}{dx} = \frac{7e^{7x} - 12x}{3y^2}$$

Example 4
Implicitly differentiate the function

$$4i^2 = 5t$$

Assume that *i* is the dependent variable and *t* is the independent variable.

Solution We approach this problem in the same way that we did all of our previous examples. Since *i* is the dependent variable and *t* is the independent variable, instead of attaching *dy/dx*, we will attach *di/dt* to any term in which we take the derivative of an *i* with respect to *t*.

Thus,

$$4i^2 = 5t$$

$$8i \frac{di}{dt} = 5$$

Executing the algebra necessary to isolate *di/dt*, we get

$$8i \frac{di}{dt} = 5$$

$$\frac{di}{dt} = \frac{5}{8i}$$

▸ Conclusion

Implicit differentiation is a tool that can be used to find the derivatives of functions whose exact form is unknown. This technique is particularly important when trying to find the rate of change of physical quantities such as displacement, or energy with respect to time. In many physical applications, we do not know the exact form of the time-dependent equation whose derivative we need to calculate. Implicit differentiation provides a method of finding this type of derivative.

Exercises with Solutions

Implicitly differentiate the following expressions:

1. $y^2 = 6x$

2. $y^2 + 3x^5 = 4x$

3. $y^3 + y^2 = 7x^4 - 6x$

4. $4y^6 - 3x^2 = 5y^2 + 9x$

5. $3y^4 + \sin(5x) = 7$

Implicitly differentiate the following functions using the designated dependent and independent variables:

6. $i^2 = 5t$ — i is the dependent variable; t is the independent variable.

7. $q^3 + 2t = 4t^3$ — q is the dependent variable; t is the independent variable.

8. $x^3 = e^{9t}$ — x is the dependent variable; t is the independent variable.

9. $p^2 + \cos(2x) = 6x^3$ — p is the dependent variable; x is the independent variable.

10. $D^4 + \sqrt{R} = \dfrac{8}{R^2}$ — D is the dependent variable; R is the independent variable.

Practice Exercises

Implicitly differentiate the following expressions:

1. $y^2 = 2x$

2. $e^{6y} = \sin(5x)$

3. $y^3 + x^2 = 8$

4. $5y^3 + \sqrt{y} + x^4 = (6x^3 + 9x^2)^7$

5. $y^5 = x^3 + e^{9x}$

6. $y^3 + e^{10y^2} + 5y^4 = \sqrt{x}$

7. $y^3 + y^2 = 16x$

8. $\dfrac{5}{y^2} + 6y^3 = \cos(x^2 + 6x - 8)$

9. $y^4 + x^2 = y^3 + 10x^5$

10. $x^2 y^3 = \sin(7x)$

Implicitly differentiate the following functions using the designated dependent and independent variables:

11. $x^2 = 5t$ — t is the independent variable; x is the dependent variable.

12. $E^3 + \cos(8t^2) = 6t - 3$ — t is the independent variable; E is the dependent variable.

13. $B^4 + B^3 = e^{7x} + \sqrt{x}$ — x is the independent variable; B is the dependent variable.

14. $\dfrac{5}{R^3} + 9R = \cos(6x)$ — x is the independent variable; R is the dependent variable.

15. $t^3 + v^4 - v = e^{5t}$ — t is the independent variable; v is the dependent variable.

DIFFERENTIAL CALCULUS

7

Maxima, Minima, and Points of Inflection

In this section we draw upon one of the two interpretations of the derivative introduced in Differential Calculus-2 to develop a method of finding points along the function at which the function is achieving extreme values. In that section, the two interpretations were given as

1. the rate at which the dependent variable is changing with respect to the independent variable.
2. the slope of the tangent line to the function at any point along the function.

It is the second of our interpretations that we will use to develop our technique for finding the extreme values of a function.

▶ Finding the Extreme Values of a Function Using the First Derivative Test

To begin, suppose that we have the following generic function:

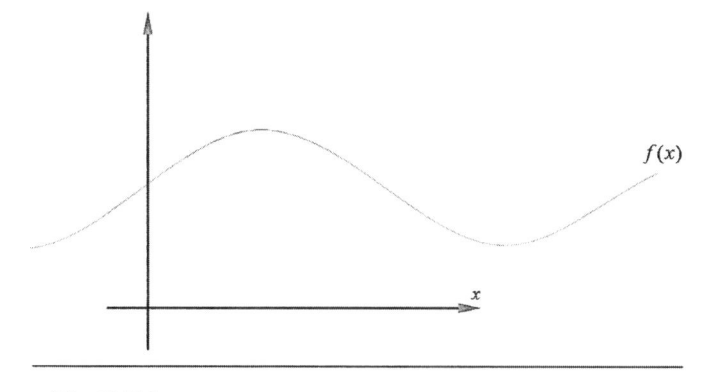

Fig. D.7.1

As we discussed in previous sections, we know that there are an infinite number of points along the curve at which we could draw a tangent line. Notice, however, that the tangent lines drawn at points 1 and 2 are horizontal and that at each of the points the function is achieving either a maximum or minimum value for the function:

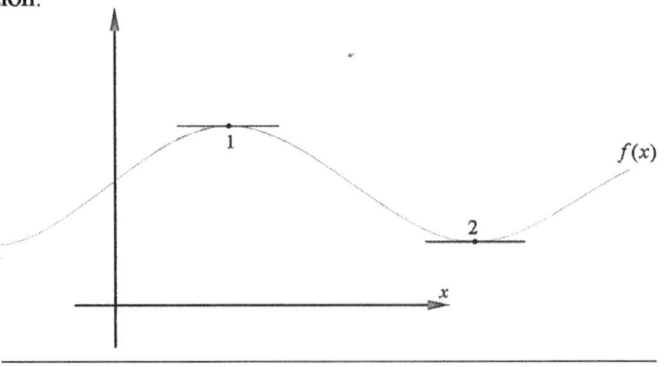

Fig. D.7.2

Because the tangent line at both points is horizontal, both of the tangent lines must have a slope of zero. Since one interpretation of the derivative of a function is as the slope of the tangent line at a point, the derivative of the function at points 1 and 2 must be equal to zero.

Reversing our logic, if we find the points along the function where the derivative has a value of zero, the tangent lines at these points must be horizontal, telling us that the function is achieving extreme values at these points. In other words, to find those points along the function at which the function is achieving either maximum or minimum values, we need only find the points along the function where the derivative has a value of zero. Because we use the first derivative of the function to determine where the function is achieving extreme values, this technique is often referred to as the *First Derivative Test*.

Example 1

Find any points on the function $f(x) = x^2$ at which the function is achieving an extreme value.

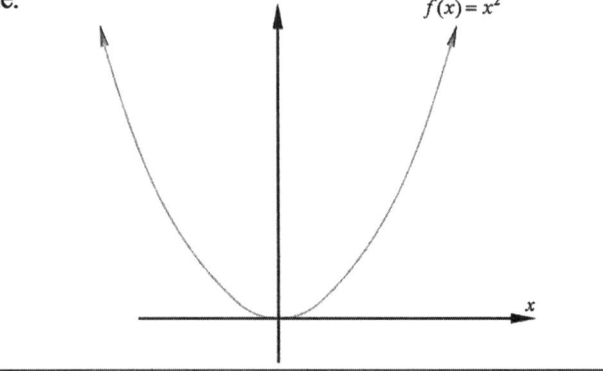

Fig. D.7.3

Solution First, we find the derivative of the function $f(x)$:

$$f(x) = x^2$$
$$f'(x) = 2x$$

To find the point(s) where the derivative has a value of zero, we set the derivative equal to zero and solve the equation for x:

$$2x = 0$$
$$x = 0$$

Therefore, the function is achieving an extreme value at the point $x = 0$. This extreme value at $x=0$ corresponds to the fact that the parabola $f(x) = x^2$ achieves its lowest value at this point:

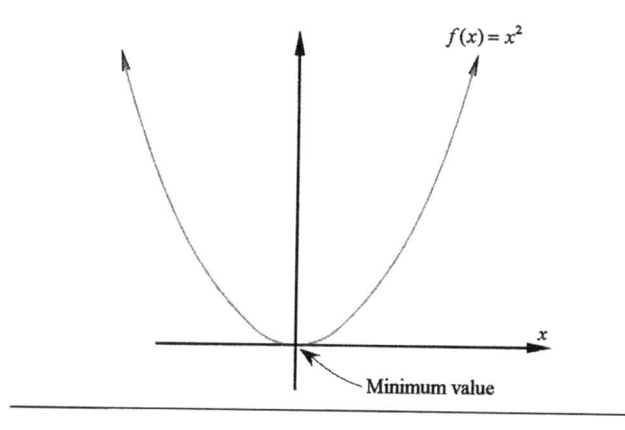

Fig. D.7.4

Notice that although our strategy produced the point at which the function achieves an extreme value, it did not tell us explicitly whether a minimum or maximum value was being achieved at this point. Thus, our next step is to develop a method by which we can determine whether a minimum or maximum value is being achieved at the point(s) produced by our technique using the first derivative.

▶ Finding Maxima and Minima Using the Second Derivative Test

To develop a method for distinguishing between maximum and minimum values on the function, we return to our sample function and points 1 and 2.

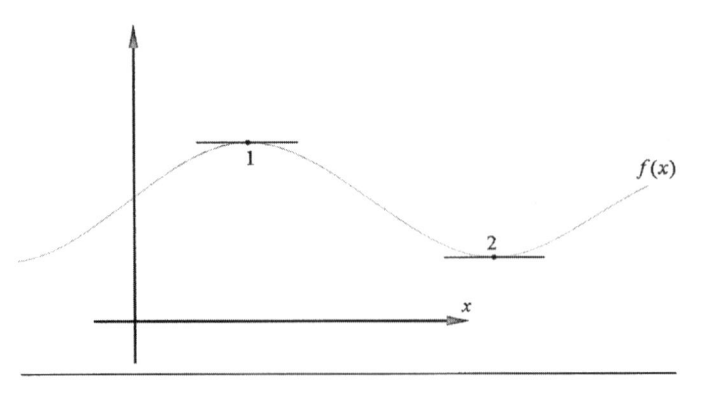

Fig. D.7.5

Notice that if we draw the tangent lines at a few points to the right of point 1, the tangent lines drawn at these points slope downward.

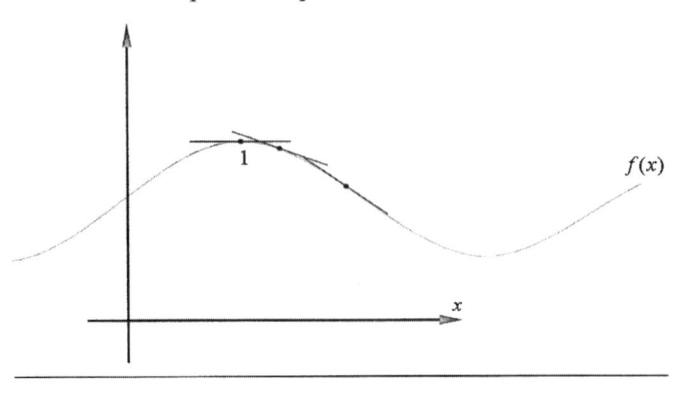

Fig. D.7.6

Notice that the slope of the tangent line at point 1 is zero, and the slopes of the tangent lines at the sample points to the right of point 1 are negative. In other words, the slopes of the tangent lines are becoming more negative as we move to the right from point 1.

If the rate of change of the function is the first derivative of the function, we can interpret the rate of change of the *slope* as the second derivative of the function. Thus, if the second derivative of the function is negative at our point of interest, the function must be achieving a maximum value at this point.

Focusing our attention now on point 2, we see that the slopes of the tangent lines drawn at points to the right of point 2 are positive.

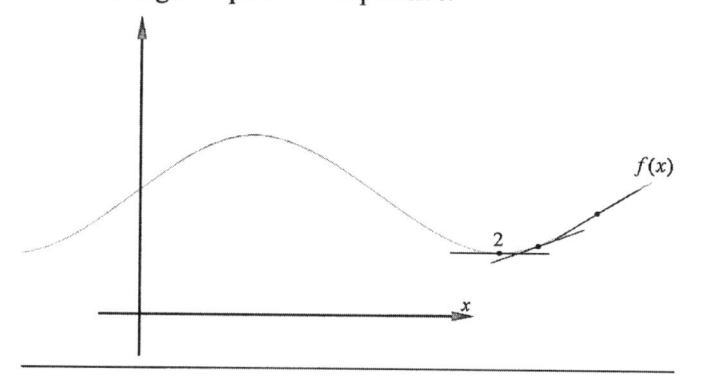

Fig. D.7.7

Since the slope of the tangent line at point 2 is zero, the rate of change of the slope as we move to the right is positive. Again phrasing this rate of change using the language of derivatives, if the second derivative of the function at the point of interest is positive, the function must be achieving a minimum value at this point.

Because we can determine whether the function is achieving a maximum or a minimum value using the second derivative of the function, this technique is referred to as the *Second Derivative Test*.

Example 2

Find any maximum or minimum values achieved by the function

$$f(x) = \frac{1}{3}x^3 - \frac{1}{2}x^2 - 6x + 5$$

on the interval (-4, 4).

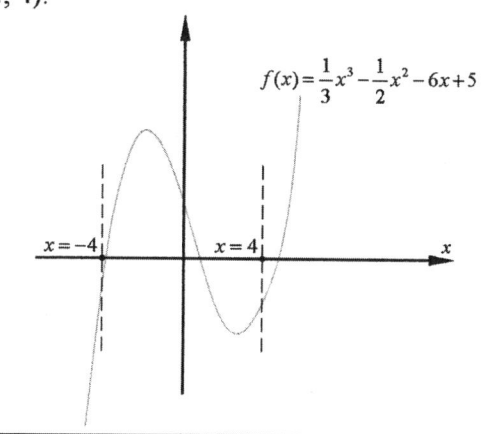

Fig. D.7.8

Solution First, we use the First Derivative Test to find the points on the function where the function achieves an extreme value. We begin by calculating the first derivative of the function:

$$f(x) = \frac{1}{3}x^3 - \frac{1}{2}x^2 - 6x + 5$$
$$f'(x) = x^2 - x - 6$$

Next, we set the first derivative equal to zero:

$$x^2 - x - 6 = 0 \qquad (1)$$

Equation (1) can be solved using factoring.

If

$$x^2 - x - 6 = 0$$

then

$$(x-3)(x+2) = 0$$

Since we are multiplying two terms and the result is zero, one of the terms must be zero.

Consequently,

$$x - 3 = 0 \quad \text{or} \quad x + 2 = 0$$

Solving each of these equations tells us that the function is achieving an extreme value at $x = 3$ and at $x = -2$. Note that each of these values of x is within the given interval of x's for this example, $(-4, 4)$.

To decide whether a maximum or minimum value is occurring at each of these points, we now use the Second Derivative Test. Once we calculate the second derivative of the function, we will evaluate it at each of our points of interest. If the second derivative is positive at the point of interest, a minimum value is occurring at the point. Conversely, if the second derivative is negative at the point of interest, a maximum value is occurring at the point.

Calculating the second derivative of the function, we find

$$f(x) = \frac{1}{3}x^3 - \frac{1}{2}x^2 - 6x + 5$$
$$f'(x) = x^2 - x - 6$$
$$f''(x) = 2x - 1$$

Next, we evaluate the second derivative at each of our two points of interest:

At $x = 3$:

$$f''(x) = 2x - 1$$
$$f''(3) = 2(3) - 1 = 5$$

Since the value of the second derivative at $x = 3$ is positive, the function achieves a minimum value at $x = 3$.

At $x = -2$:

$$f''(x) = 2x - 1$$
$$f''(3) = 2(-2) - 1 = -5$$

Since the value of the second derivative at $x = -2$ is negative, the function achieves a maximum value at $x = -2$.

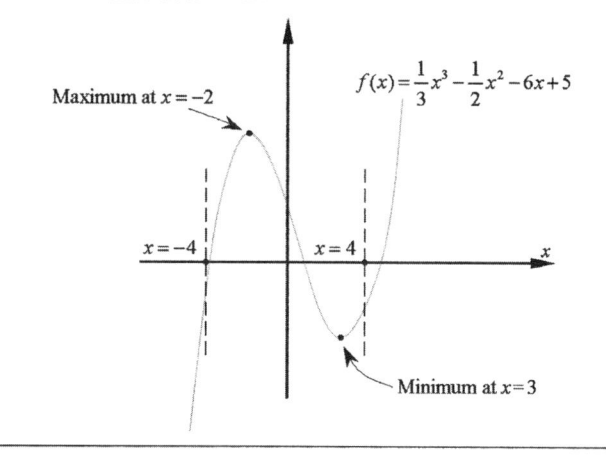

Fig. D.7.9

Example 3

Find any maxima or minima for the function $f(x) = x^2 + 1$.

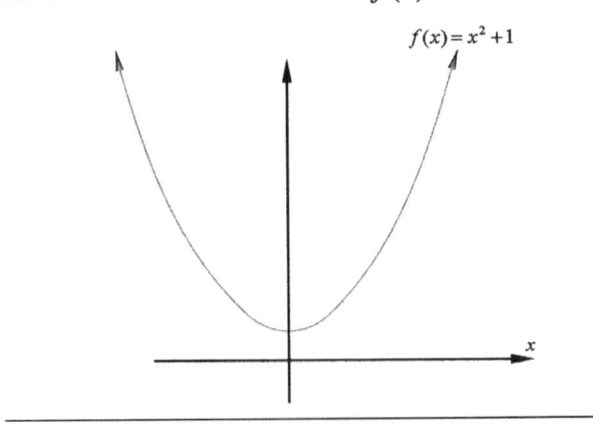

Fig. D.7.10

Solution First, we use the First Derivative Test to find any points on the function where extreme values occur. Finding the first derivative of the function give us

$$f(x) = x^2 + 1$$
$$f'(x) = 2x$$

Next, we set the first derivative equal to zero and solve for x:

$$2x = 0$$
$$x = 0$$

The function achieves an extreme value at $x = 0$. To find whether a minimum or maximum value occurs at this point, we apply the Second Derivative Test.

First, we find the second derivative:

$$f(x) = x^2 + 1$$
$$f'(x) = 2x$$
$$f''(x) = 2$$

Notice that in this example, the second derivative does not provide an x into which we can substitute our point of interest $x = 0$. However, we can still use the second derivative to decide whether a minimum or maximum value is occurring.

Since the second derivative is positive $(f''(x)=2)$, a minimum value for the function occurs at the point of interest, $x = 0$.

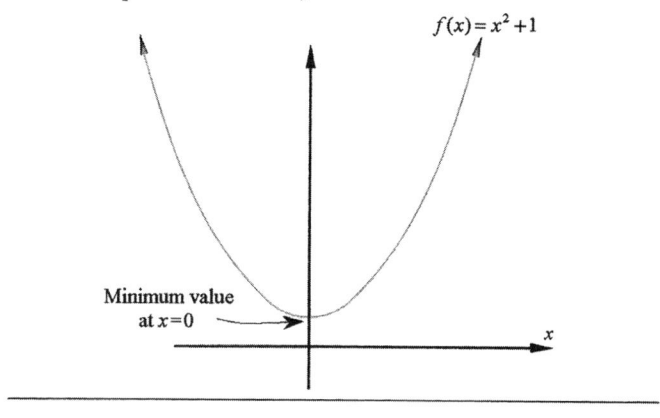

Fig. D.7.11

▶ Absolute Versus Local Maximum and Minimum Values

In this section, our purpose is to compare the language of the problem statements used in Examples 2 and 3 to gain a more precise understanding of the concepts of maxima and minima. If we return to each of these problem statements, we see that in Example 2 we were careful to specify a region of the function that was being analyzed, while in Example 3 we did not include this type of specification. This distinction can be understood by again looking at the graph of the function in Example 3:

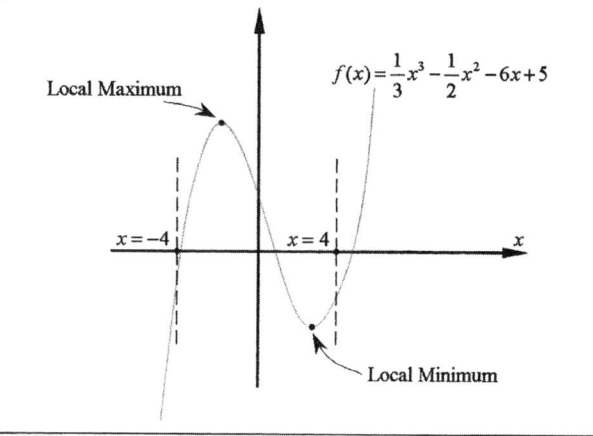

Fig. D.7.12

Although the function achieved a *local* maximum value at $x = -2$, it cannot be the *absolute* maximum value for the function since the function increases indefinitely as we move along the positive x-axis from $x = 3$. Consequently, although the function

$$f(x) = \frac{1}{3}x^3 - \frac{1}{2}x^2 - 6x + 5$$

has a local maximum at the point $x = -2$, it does not have an absolute maximum.

Similarly, if we look at the minimum value achieved at $x = 3$, this minimum must be a *local* minimum, since the function decreases indefinitely as we move to the left from $x = -2$ on the negative x-axis. Consequently, the function has a local minimum occurring at $x = 3$, but no absolute minimum.

If we now look at the function from Example 3, we see that the minimum occurring at $x = 0$ is indeed an absolute minimum, since the function increases indefinitely on both side of this point:

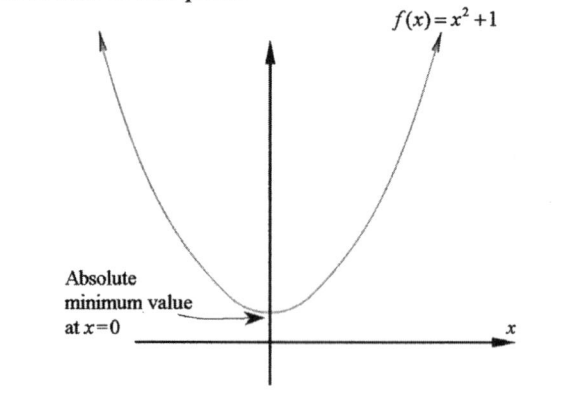

Fig. D.7.13

Because the minimum value was absolute, we did not need to specify in Example 3 a region in which we were analyzing the maximum and minimum values.

▸ Points of Inflection

In closing this chapter, we consider one more result produced by the Second Derivative Test. In addition to finding any maximum or minimum values that a function may be achieving, it also allows us to discuss the curvature of the function at various points along the function.

Consider the following two graphs:

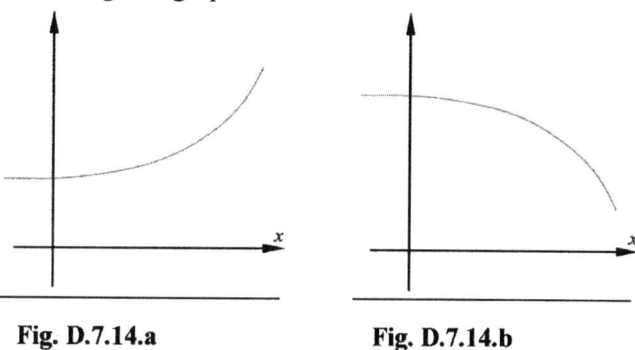

Fig. D.7.14.a   **Fig. D.7.14.b**

Notice that in Figure D.7.15.a the function lies above the tangent line drawn at point x, while in Figure D.7.15.b the function lies below the tangent line drawn at the same point:

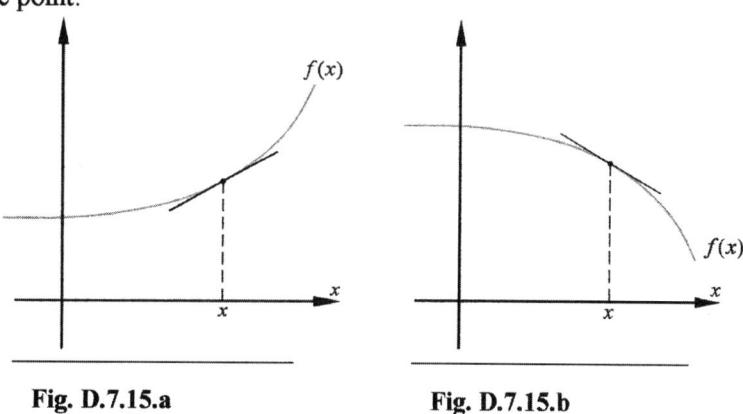

Fig. D.7.15.a   **Fig. D.7.15.b**

In addition to classifying the function as to whether it lies above or below its tangent line, we can discuss its type of curvature, or *concavity*. We say that the function in Figure D.7.15.a is *concave upward*, and the function in Figure D.7.15.b is *concave downward*.

Since we know that the curvature and tangent lines of a function are related to the second derivative, we can express the concavity of a function at a point in terms of the second derivative.

The Second Derivative and Concavity

Suppose that the second derivative of a function exists on an interval I of the function.

1. If $f''(x) > 0$ on the interval, the function is *concave upward* on the interval.
2. If $f''(x) < 0$ on the interval, the function is *concave downward* on the interval.

Points of Inflection

Up to this point, we have analyzed situations in which the sign on the second derivative was either positive or negative. We see that this sign can provide information about any maxima or minima that may be occurring as well as predict the concavity of the function. Now, let's consider the case in which the second derivative of the function is *equal* to zero.

In Figure D.7.16, the function is changing its concavity at the point x:

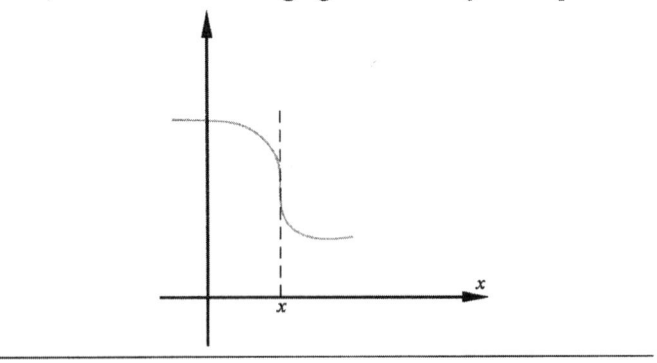

Fig. D.7.16

If the concavity on one side of a point is concave upward and on the other side the concavity is concave downward, the point is called a *point of inflection*.

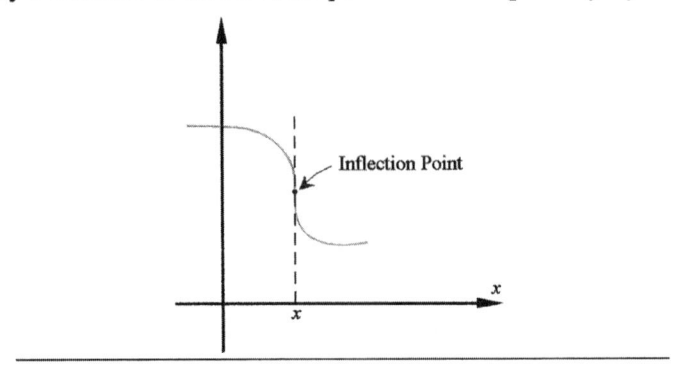

Fig. D.7.17

Possible points of inflection can be found using the Second Derivative Test. Because the concavity of a function changes at a point of inflection, the sign on the second derivative must be different on either side of the point. At a point of inflection, $f''(x) > 0$ on one side of the point, and $f''(x) < 0$ on the other side. Consequently, the second derivative must have a value of zero at the point where the concavity is changing. This means that $f''(x) = 0$ at a point of inflection.

Therefore, we can use the Second Derivative Test to identify possible inflection points on a function. If the second derivative is equal to zero at a point on the function, the point may be a point of inflection. Since the concavity of the function must change at the point in order for it to be a point of inflection, the second derivative must be evaluated at points to the right and left of the possible point of inflection. The signs on the second derivatives must be different on opposite sides of the point of interest in order for it to be a point of inflection. *Care should be taken to choose points close to the possible point of inflection since the function may be achieving a variety of local maximum and/or minimum values on an interval.*

To summarize the techniques discussed in this chapter:

The First Derivative Test

1. Find the first derivative of the function and set it equal to zero.
2. Solve the equation from Step 1 to identify points on the function at which a maximum value, a minimum value, or a point of inflection may be occurring.

The Second Derivative Test

1. Find the second derivative of the function.
2. Insert each point of interest into the second derivative.
3. The sign on the second derivative at the point of interest can be used to predict whether a maximum, minimum, or point of inflection is occurring at the point.

If $f''(x) > 0$ at the point of interest, a *local minimum* is occurring at the point. If this local minimum is the absolute smallest value that the function achieves, the point is then referred to as an *absolute minimum*.

If $f''(x) < 0$ at the point of interest, a *local maximum* is occurring at the point. If this local maximum is the absolute largest value that the function achieves, it is then referred to as an *absolute maximum*.

If $f''(x) = 0$ at the point of interest, the point may be a *point of inflection*. The signs on the second derivative at points a small distance to the right and to the left of the point should then be calculated. If the signs on the second derivates are opposite on either side of the point, the concavity of the function is changing at this point and it is indeed a point of inflection.

Exercises with Solutions

Find any/all local maxima, local minima, and points of inflection for the following functions:

1. $f(x) = 3x^2$

2. $f(x) = -x^2$

3. $f(x) = \frac{1}{3}x^3 - \frac{1}{2}x^2 - 12x + 6$

4. $f(x) = 8x^2 - 1$

5. $f(x) = x^3$

Practice Exercises

Find any/all local maxima, local minima, and points of inflection for the following functions. Assume an interval of $(-\infty, \infty)$ unless otherwise indicated.

1. $f(x) = \frac{1}{3}x^3 - x + 9$

2. $f(x) = x^2 + 3x$

3. $f(x) = (x+1)^2$

4. $f(x) = 3x(x-1)$

5. $f(x) = 10x + \frac{3}{2}x^2 - \frac{1}{3}x^3$

6. $f(x) = 5x^4$

7. $f(x) = \dfrac{x^2 + x}{2} - 6$

8. $f(x) = \tan x$ on the interval $\left(-\dfrac{\pi}{2}, \dfrac{\pi}{2}\right)$

9. $f(x) = \sin x$ on the interval $(0, \pi)$

10. $f(x) = \cos x$ on the interval $(0, 2\pi)$

Differential Calculus

Applications

1. The velocity, v, of an object is the rate of change of the displacement, s, with respect to time,

$$v = \frac{ds}{dt}$$

   and the acceleration, a, is the rate of change of the velocity

$$a = \frac{dv}{dt}$$

   Problem: Given that a certain object has a displacement equation of

$$s = t^3 + 3t^2$$

   find the displacement, velocity, and acceleration of the object at $t = 1$ sec.

2. For conservative physical systems, the potential energy of a system, PE, and the force on the system, F, can be related using differential calculus. The force is simply the negative rate of change of the potential energy with respect to the displacement:

$$F = -\frac{d(PE)}{dx}$$

   Problem: Given this relationship between force and potential energy, find an expression for the force for each of the following systems given its potential energy equation:

   a) $PE = \frac{1}{2}x^2$

   b) $PE = e^{-5x}$

   c) $PE = \ln x + \sin x$

104 Differential Calculus

3. Simple harmonic systems are physical systems that execute periodic motion. In other words, they return to their original locations with a regular frequency. Because of this periodic motion, their displacement equations contain trigonometric functions rather than simple polynomials.

    Problem: For the simple harmonic oscillator given by

    $$x = 5\cos(2t)$$

    find the displacement, velocity, and acceleration of the oscillator at $t = 0$ sec and $t = 1$ sec. (*Hint:* Review Differential Calculus Application 1).

4. The force on a moving mass is related to the momentum, p, of the mass according to

    $$F = \frac{dp}{dt}$$

    Problem:

    a) Given the above relationship between force and momentum, with respect to what independent variable is the momentum changing?

    b) If the momentum of a particular mass is changing according to the equation

    $$p = \sqrt{t} + e^t$$

    find an expression for the time-dependent force acting on the mass.

5. The amount of electric current, i, that flows through a region of a circuit is defined to be the amount of electric charge, q, that passes through the region in a certain period of time. Expressed in the language of calculus, the current flow is the time-rate of change of the electric charge:

    $$i = \frac{dq}{dt}$$

    Problem: Find an expression for the time-dependent current flow given each of the following time-dependent charge equations:

    a) $q = 6t^3$

    b) $q = e^{-\frac{t}{5}}$

    c) $q = \sin 5t$

6. An RC-circuit contains a direct current voltage source, *V*, a resistor, *R*, and a capacitor, *C*, as in the following figure:

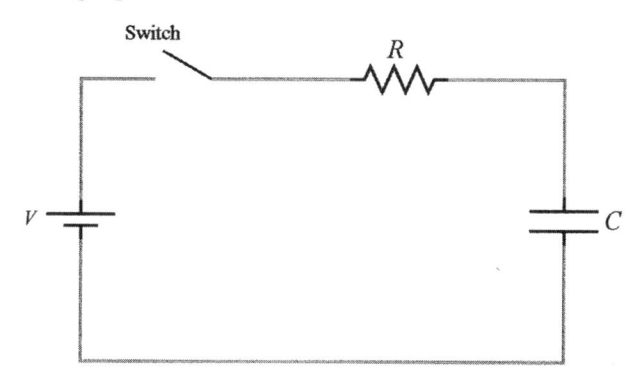

When the switch in the circuit is closed, the capacitor will charge according to the equation:

$$q = CV\left(1 - e^{-\frac{t}{RC}}\right)$$

Problem:

  a) Find an equation that will yield the current flow during the charging process at any time, *t*. (*Hint:* See Differential Calculus Application 5).
  b) Find the amount of current flowing through an RC-circuit that contains a 10 volt source, a 5000 ohm resistor, and a 6 x 10^{-6} farad capacitor at $t = 0.005$ seconds.

7. A transverse wave traveling along a rope can be described by the equation

$$y = A\sin(kx - \omega t)$$

where

- *y* is the location of a particular rope element, above or below equilibrium, at the time *t*.

- *A*, the amplitude of the wave, is defined as the maximum displacement of each rope element above or below equilibrium as the wave passes.

- *k* is the angular wave number. If we construct a graph using radians along the horizontal axis, *k* is the number of full waves that would fit in 2π radians. It is related to the wavelength, λ, by

$$k = \frac{2\pi}{\lambda}$$

- x is the horizontal location of the rope element; it tells us how far along the rope the rope element is located.

- ω is the angular frequency of the wave, measured in radians per second. It is related to the linear frequency, measured in hertz, by

$$\omega = 2\pi f$$

- t is the time of interest of the location of the rope element at x.

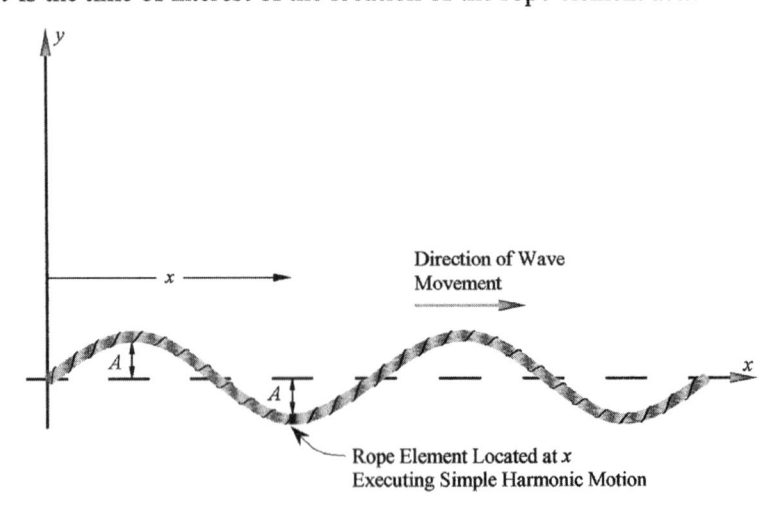

Problem: For the wave given by the equation

$$y = 10\sin(5x - 4t)$$

find

a) the rate of change of the wave equation with respect to the displacement, x.

b) the rate of change of the wave equation with respect to time.

8. The amount of a radioactive sample that remains after a time, t, is given by the equation

$$N = N_0 e^{-\lambda t}$$

where

- N is the amount of the sample left after a time t.

- N_0 is the original size of the sample at the time $t = 0$.

- λ is the decay constant of the material.

- t is the time at which the sample is being analyzed.

Problem:

- a) For a sample that has a decay constant of 100, find the amount of time required until only ½ of the original sample remains.
- b) Find a general expression for the rate of change of a sample size with respect to time.

9. An LC-circuit contains an inductor, L, and a charged capacitor, C, as in the following figure:

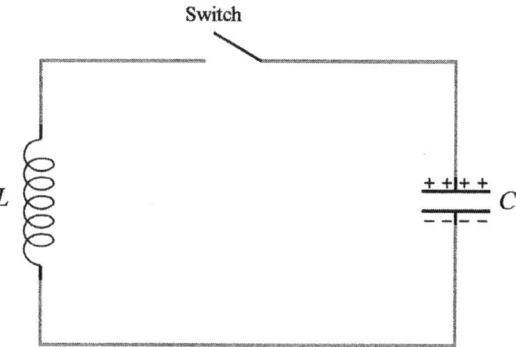

When the switch is closed in the circuit, the current will alternate with a frequency determined by both the inductor and the capacitor. Although the amount of electric charge, q, stored on the capacitor depends upon when we look at it and is therefore time-dependent, the overall amount of *energy* in the circuit is *time-independent*. Expressed in the language of calculus:

$$\frac{d}{dt}(Energy) = 0$$

Given that the amount of energy stored in a capacitor, C, with a charge of q on its plates is given by

$$E_{Cap} = \frac{q^2}{2C}$$

and the amount of energy stored in an inductor, L, with a current of i flowing through it is given by

$$E_{Ind} = \frac{1}{2}Li^2$$

the derivative expression

$$\frac{d}{dt}(Energy) = 0$$

becomes

$$\frac{d}{dt}(E_{Ind} + E_{Cap}) = 0$$

$$\frac{d}{dt}\left(\frac{1}{2}Li^2 + \frac{q^2}{2C}\right) = 0 \quad \textbf{(E1)}$$

Since the amount of charge, q, and the current flow, i, both depend upon the time at which we analyze the circuit, each is a function of the variable t:

$$q = q(t)$$
$$i = i(t)$$

Problem: Using implicit differentiation, show that Equation (E1) becomes

$$L\frac{d^2q}{dt^2} + \frac{q}{C} = 0$$

10. When analyzing revolving systems, it is often easier to express the motion of the system using angular rather then linear variables. For example, the angular variable theta (θ) is used to represent the angular distance traveled. Accordingly, it is possible to refer to the amount of angle through which a system rotates in a certain period of time as the *angular velocity* rather than the linear velocity. This angular velocity is usually assigned the Greek letter omega - ω. Since the angular velocity is the time-rate of change of the angular displacement, we can express the relationship between them as

$$\omega = \frac{d\theta}{dt}$$

Similarly, we can express the *angular acceleration*, α, as the rate of change of the angular velocity:

$$\alpha = \frac{d\omega}{dt}$$

or as the second derivative of the angular displacement:

$$\alpha = \frac{d^2\theta}{dt^2}$$

Problem: Given that the angular displacement for a certain system is

$$\theta = 2t^3 + 5t^2 + t$$

find the angular displacement, velocity, and acceleration at $t = 2$ sec. Assume that the angular displacement is measured in units of radians.

11. The population growth in a certain city is given by the equation

$$P = 100e^{2t}$$

where t is expressed in years.

Problem: Find the rate of change of the population in a two-year period.

12. The current in a certain region of a circuit is given by the time-dependent equation

$$i = \frac{1}{3}t^3 - \frac{1}{2}t^2 - 6t$$

Problem: Find any/all times at which the current flow in the circuit achieves a maximum or minimum value. Identify whether the time corresponds to a maximum or minimum value of the current flow.

13. The magnitude of the magnetic field B that surrounds a straight current-carrying wire is inversely proportional to the distance r from the wire that the field is measured.

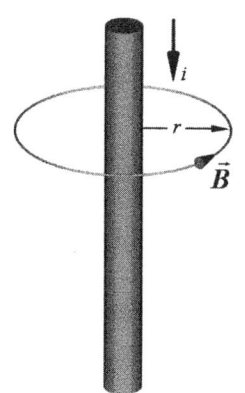

110 Differential Calculus

The magnitude of the magnetic field can be found from the equation

$$B = \frac{\mu_0 i}{2\pi r}$$

where the constant μ_0 is a physical constant known as the *permeability*,

$$\mu_0 = 4\pi \times 10^{-7} \text{ Tm/A}$$

and i is the current flow measure in amps.

Problem: A long, thin, straight wire has a current of 0.005 amps moving through it. Find the rate of change of the magnetic field with respect to r if we make the measurement at $r = 0.001$ m.

14. The total current through a pn-junction can be found using Shockley's Equation:

$$I_T = I_S \left(e^{\frac{Vq}{kT}} - 1 \right)$$

where

- I_S = the reverse saturation current through the diode
- e = the base of the natural logarithm
- q = the magnitude of the charge on the electron = 1.6×10^{-19} C
- V = the voltage across the depletion layer
- T = the temperature of the device, expressed in Kelvin (K)
- k = Boltzmann's constant = 1.38×10^{-23} J/K

Problem: If we assume that only the voltage V in the equation is a variable, find an expression for the rate of change of the total current flow with respect to the voltage across the depletion layer.

15. The velocity at which the fluid in a container exits the container is determined by Torricelli's Equation

$$v = \sqrt{2gh}$$

where h is the height of the fluid above the aperture and g is the acceleration due to gravity, 9.8 m/s^2:

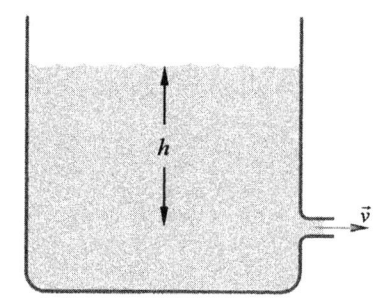

Thus, the velocity of the fluid is proportional to the height of the fluid and decreases as the fluid level decreases.

Problem: Find the rate of change of the fluid velocity when the fluid level has fallen to a height of 0.25 m above the aperture.

Integral Calculus

- Integration Preliminaries

- Finding the Integral of a Function
    - The Antiderivative of a Function
    - Using the Antiderivative to Calculate the Integral of a Function
    - Definite and Indefinite Integrals

    Exercises with Solutions
    Practice Exercises

- The Method of Substitution

    Exercises with Solutions
    Practice Exercises

- Differentials II – The Relationship between Integral and Differential Calculus

    Exercises with Solutions
    Practice Exercises

- Integration by Parts

    Exercises with Solutions
    Practice Exercises

- Numerical Integration Techniques

    - Endpoint Approximations
    - The Left-endpoint Approximation
    - The Right-endpoint Approximation
    - The Trapezoidal Approximation

    Exercises with Solutions
    Practice Exercise

- Integral Calculus Applications

INTEGRAL CALCULUS

1

Integration Preliminaries

In Differential Calculus 2, we discussed a fundamental definition for the derivative of a function, and then in Differential Calculus 3 we learned rules that will generate the derivatives of functions quickly. In this section, we will take the same approach. After an introduction to the background information necessary for an understanding of *integration*, we will move to algebraic methods for calculating the *integral* of a function.

Suppose we have a rectangle that has a height of 6 and a width of 4:

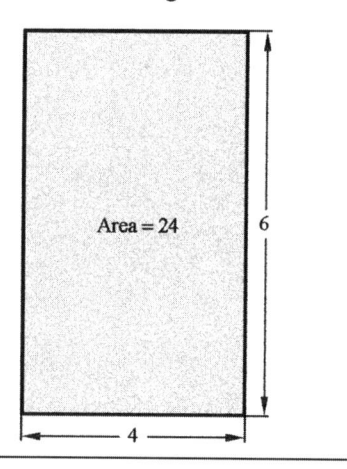

Fig. I.1.1

We can find the area of the rectangle by multiplying its height by its width:

$$\text{Area} = \text{height} \times \text{width}$$
$$\text{Area} = 6 \times 4 \text{ sq. units}$$

It is possible, however, to find the area of the rectangle in a slightly different way. Suppose that first we break the rectangle into four smaller rectangles, each of which has a width of 1:

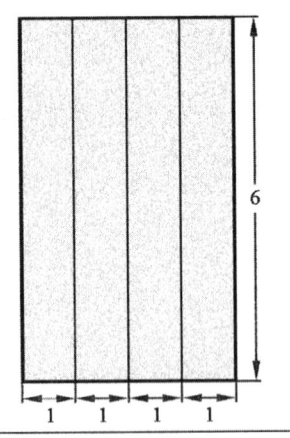

Fig. I.1.2

As we did with the large rectangle, we can find the areas of each of the smaller rectangles by multiplying their height by their width. If we carry out these multiplications, we see that each of the smaller rectangles has an area of 6:

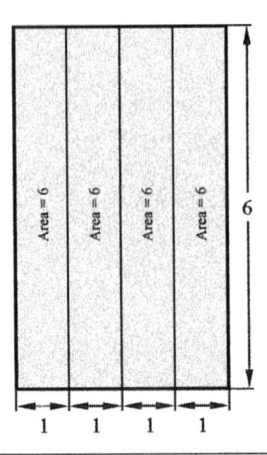

Fig. I.1.3

We can find the area of our original rectangle by adding together the areas of each of the smaller rectangles:

$$\text{Area of large rectangle} = 6 + 6 + 6 + 6 = 24 \text{ sq. units}$$

The concept of finding the area of a geometric figure by breaking it into a sum of smaller areas is central to the theory of integration.

▸ The Integral of a Function

Suppose we have the following function, and that we are interested in the area that is under the curve between points *a* and *b*:

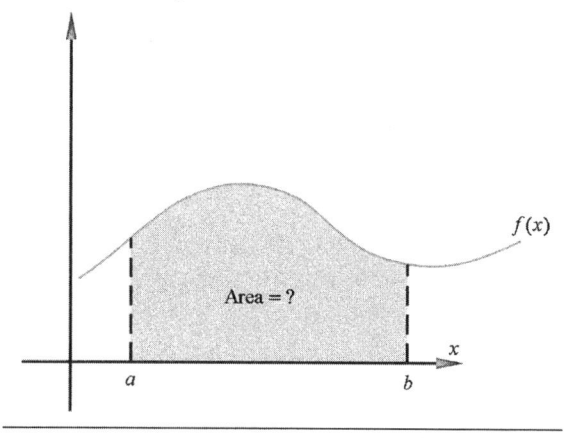

Fig. I.1.4

Unfortunately, this is not a simple geometric shape that has a simple "plug and chug" equation that we can use to find the area. We can, however, use the strategy of calculating an area by using a sum of smaller areas to develop a technique for finding the area under this function. Just as we did with the rectangle in Figure I.1.1, let's take the area under this curve and fill it with a series of rectangles with *identical widths*. Since the width of each rectangle is along the *x*-axis, we will identify the width of each rectangle as Δx. Now that we know the widths of our rectangles, we must decide on a method of describing their heights.

To choose the height of each rectangle, we start each on the *x*-axis and have it rise until its left corner touches the curve:

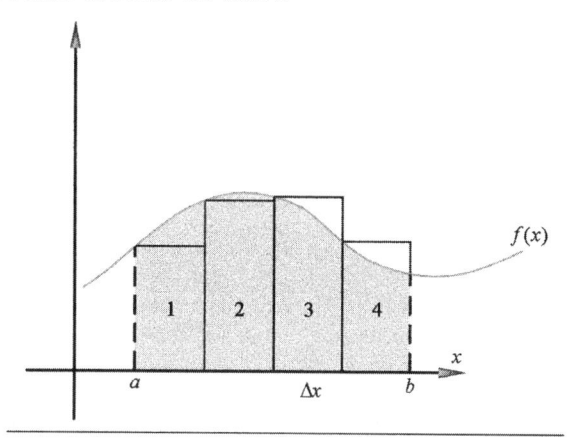

Fig. I.1.5

Notice that each rectangle whose left edge is located at a point on the x-axis has a height determined by $f(x)$ since each rectangle rises up until it intersects the function $f(x)$.

Since we now have expressions for both the height and the width of our rectangles, we can express their areas by multiplying the height by the width. For example, the area of rectangle 1 in Figure I.1.5 can be expressed as

$$\text{Area}_1 = f(x)\Delta x$$

We can now get an approximation of the area under the curve between the points a and b by adding together the areas of the four rectangles. Using the capital Greek letter sigma, Σ, which means sum, we can express the approximate area as

$$\text{Area} = \sum_{a}^{b} f(x)\Delta x$$

Notice that in the area expression we have the letters b and a above and below the summation symbol. This notation means that we are to begin adding the areas of the rectangles at the point a and stop adding at the point b.

Returning to Figure I.1.5, we see that although our technique gave us an approximation of the area under the curve, it was not a very good one. There are regions of area that were not included in any rectangle. Also, there are rectangles that contain area that does not lie under the curve:

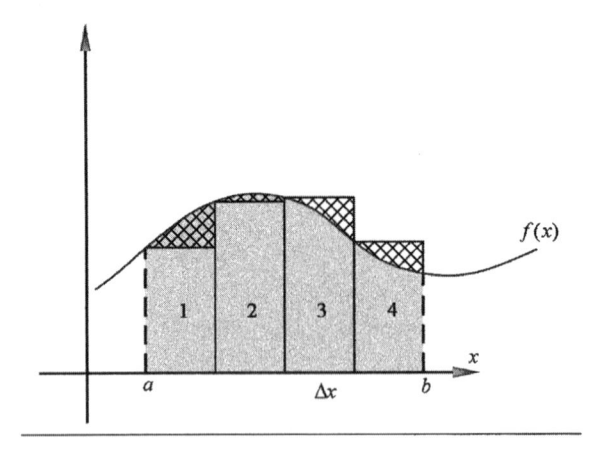

Fig. I.1.6

Suppose, however, that instead of inserting four rectangles of equal widths into the area under the curve, we insert eight. We will use the same method as before and assign a width of Δx and a height of $f(x)$ to each of the rectangles:

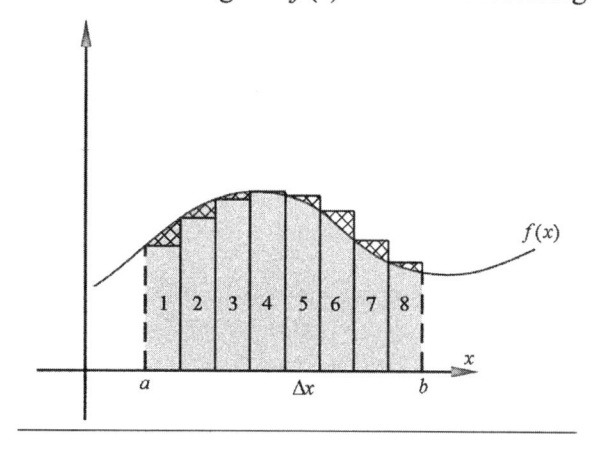

Fig. I.1.7

We now have a better approximation of the area under the curve. We can continue this process as long as we wish. Instead of inserting eight rectangles, we insert 16, 32, 64, etc. Each time we increase the number of rectangles inside the area, we decrease their widths. Narrower rectangles lead to a better approximation of the area under the curve. We miss less and less area, and we do not take into account area that does not lie under the curve.

Now suppose we take a drastic step and allow our rectangles to become infinitely narrow. Or, using the language of mathematics used in Differential Calculus 1, let's take the *limit* as Δx approaches zero. A mental picture of such a maneuver would have rectangles that looked like straight vertical lines. In this case, we will not miss any area or take into account any area that does not actually lie under the curve. In other words, we will have the actual area under the curve between points a and b:

$$\text{Area} = \lim_{\Delta x \to 0} \sum_{a}^{b} f(x) \Delta x$$

At this point, it helps to introduce a slight change in notation. We will express the widths of our infinitely narrow rectangles differently from the widths of our normal rectangles. Up until this point, we have been referring to the widths as Δx. Remember that the symbol Δ is the Greek letter *delta*. Let's take the first letter of the word *delta*, d, and express the widths of our infinitely narrow rectangles as dx. This will distinguish for us between those rectangles that have a regular thickness (Δx) and those that have an infinitely small thickness (dx).

Using this new notation, we can express the area of one of our infinitely small rectangles as

$$\text{Area}_{\text{infinitely narrow}} = f(x)dx$$

However, the fact that our rectangles are now infinitely narrow forces us to consider another point. Each time we reduced the widths of our rectangles, we increased the number of them that was required to fill the area under the curve. If our rectangles are now infinitely narrow, we must add up an *infinite* number of them in order to find the area under the curve between a and b.

Just as we changed the notation for the widths of our rectangles to remind us that they are now infinitely narrow, we are going to change our summation notation. To remind us that we are adding up an infinite number of infinitely narrow rectangles, we will replace the summation symbol Σ. Instead, we will use a Gothic letter S (the first letter of the word *summation*).

In our definition of area, we will replace

$$\lim_{\Delta x \to 0} \Sigma$$

with

$$\int$$

keeping the limits a and b the same. This stretched out S is normally referred to as the *integration symbol*. It reminds us that we are adding up an infinite number of items. Using these two changes in notation, we can rewrite our area expression

$$\text{Area} = \lim_{\Delta x \to 0} \sum_{a}^{b} f(x) \Delta x$$

as

$$\text{Area} = \int_{a}^{b} f(x) dx$$

> The area under a function between two endpoints is called the *integral of the function* between those two points. When we "integrate" the function $f(x)$ between points a and b, we are finding the area under the function between them.

Before we move on to how actually to calculate an integral, we close this section with a summary of the various pieces of the integral expression.

Parts of the Integral Expression

1. dx is the width of one of our infinitely narrow rectangles.

2. $f(x)$ is the height of each of our infinitely narrow rectangles.

3. $f(x)dx$ is the area of each of our infinitely narrow rectangles.

4. a is the point at which we begin adding the areas of our infinitely narrow rectangles. This point is often referred to as the "lower limit of integration".

5. b is the point at which we stop adding the areas of our infinitely narrow rectangles. This point is often referred to as the "upper limit of integration".

6. The whole expression

$$\int_a^b f(x)dx$$

is the total area under the curve between a and b.

Now that we know that one interpretation of an integral is the area under a curve, we will now discuss how actually to calculate the integral of a function.

INTEGRAL CALCULUS

2

Finding the Integral of a Function

In Integral Calculus 1, we found that one way to interpret the integral of a function is as the area under the function. In this section, we want to develop a technique that will allow us to actually calculate an integral.

> **FYI**
>
> The area of mathematics referred to as *Integral Calculus* involves being able to find and use the integrals of functions.

Just as in the case of the derivative, there are some elaborate and mathematically difficult ways to calculate an integral. In the section on derivatives, we began by using a difficult method involving limits, and then we moved into shortcut rules that allowed us to find the derivative quickly. In this section, we move directly into the shortcut. Because our goal is to be able to use the ideas of integral calculus to solve physical problems, we will take for granted that there exist many mathematically difficult and rigorous ways of calculating integrals. As we begin learning this shortcut, it is important for us to keep in the backs of our minds that we could be using much more formal techniques to find these integrals. Our first step in learning how to integrate functions involves learning how to find the *antiderivative* of a function.

▸ The Antiderivative of a Function

Suppose we have an expression such as $2x$. If we want to find the *antiderivative* of $2x$, we are looking for an expression that, if we take its derivative, would give us $2x$. Finding the antiderivative of a function (and hence the integral of the function) is basically the reverse of finding the derivative. The function whose derivative would be $2x$ is x^2.

The derivative of x^2 is $2x$.

The *antiderivative* of $2x$ is x^2.

Example 1
Find the antiderivative of $3x^2$.

Solution Again, this is another way of asking for the function that, if we take its derivative, will give us $3x^2$. By inspection, we see that the antiderivative of $3x^2$ is x^3, since the derivative of x^3 is $3x^2$.

Unknown Constants

Now that we have a beginning understanding of the concept of an antiderivative, we need to polish it slightly. Suppose we start with the two functions

$$f(x) = x^2 + 12 \quad \text{and} \quad f(x) = x^2 + 3$$

If we now find the derivative of each of these functions

$$f(x) = x^2 + 12 \qquad\qquad f(x) = x^2 + 3$$
$$f'(x) = 2x \qquad\qquad\qquad f'(x) = 2x$$

we see that both functions have the same derivative. In other words, we can add any number to x^2 and it does not affect the derivative. Consequently, there are an infinite number of functions that have a derivative of $2x$. This means that if we would like to find the *antiderivative* of $2x$, we must allow for the possibility that there is an unknown constant added to the x^2 term. To incorporate this into our antiderivative, we add the letter C to represent the unknown constant.

Thus, the correct antiderivative for $2x$ is actually $x^2 + C$.

Example 2

Find the antiderivative of $3x + 4$.

Solution We will attack this problem one piece at a time. First, the antiderivative of $3x$ is

$$\frac{3}{2}x^2$$

Note that we can check our work by finding the derivative of our expression. If we have found the correct antiderivative of $3x$, when we take the derivative of our answer it should yield $3x$. Using our rule from Differential Calculus 3, we see that the derivative of

$$\frac{3}{2}x^2$$

is indeed $3x$.

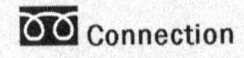

> **Connection**
>
> The rule for finding the derivatives of simple polynomials is introduced on page 38 of the Differential Calculus section.

Next, we must address the 4 term in our problem. Again, we are looking for the expression that, if we take its derivative, we will get 4. The solution is $4x$. The derivative of $4x$ is 4 and the *antiderivative* of 4 is $4x$.

Putting our two pieces together, our antiderivative thus far is

$$\frac{3}{2}x^2 + 4x$$

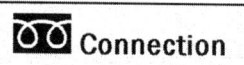

Connection

For a discussion of why the derivative of a constant is zero, see Differential Calculus 3.

The last step is for us to attach the letter C at the end to remind us that there might be an unknown constant that cancels in the process of our taking the derivative. Thus, the antiderivative of $3x + 4$ is

$$\frac{3}{2}x^2 + 4x + C$$

Finding the antiderivative of the expression in Example 2 was not too difficult. We were able to look at each term and reason out the required antiderivatives. However, we have no guarantee that the physical problems that we will confront will always have polynomials that are this simple. What we need is an algorithm for finding antiderivatives of simple polynomials that will work even if we are not able to stare at the expression and guess an answer. Here is such an algorithm.

Algorithm for Finding the Antiderivatives of Simple Polynomials

1. Take the exponent on the term and add 1 to it.
2. Take this new exponent, invert it, and make this a coefficient in front of the antiderivative.

It is possible to write the above algorithm in a formula similar to the summary table at the end of Differential Calculus 3. The antiderivative of a term of the form

$$ax^n$$

is

$$\left(\frac{1}{n+1}\right)ax^{n+1}$$

Example 3

Find the antiderivative of x^5.

Solution The first step in the algorithm is to raise our current exponent, in this case 5, by one power, yielding x^6. Next, we must find the correct coefficient to place in front of the antiderivative. The second step in the algorithm tells us that this coefficient can be found by taking our new exponent, 6, and inverting it, yielding 1/6.

Thus, the antiderivative of x^5 is

$$\frac{1}{6}x^6 + C$$

Note that we have once again added C to represent the possible unknown constant in the function.

Example 4

In this example, let's find the antiderivative of an expression that has a coefficient other than 1 in front of the term.

Find the antiderivative of $4x^2$.

Solution Our algorithm for finding the antiderivative still works even though there is now a 4 multiplying our expression. The first step is to take the exponent, in this case 2, and raise it by one power, yielding x^3. Next, we must find the correct coefficient to place in front of the term. The algorithm tells us to take the new exponent, 3, and invert it. If we invert 3, we get 1/3. We now take the 1/3 and use it as a coefficient, *remembering to multiply by the 4 that was already in front of the term.*

Consequently, the antiderivative of $4x^2$ is

$$\frac{1}{3}(4)x^3 + C$$

which can be simplified to

$$\frac{4}{3}x^3 + C$$

In these next few examples, we extend the concept of an antiderivative to expressions that are not simple polynomials. As we will see, our approach will be the same as in all of the previous examples that involved simple polynomials.

Example 5
Find the antiderivative of e^{2x}.

Solution The antiderivative of e^{2x} is

$$\frac{1}{2}e^{2x} + C$$

since the derivative of our solution reproduces e^{2x}:

$$\frac{d}{dx}\left(\frac{1}{2}e^{2x} + C\right) = \frac{1}{2} \cdot 2e^{2x} + 0 = e^{2x}$$

Notice that it was necessary to include a factor of ½ in the antiderivative to compensate for the factor of 2 that is produced from the rule for exponential functions. Note also that we have once again added the letter C to the antiderivative to represent a possible unknown constant.

Example 6
Find the antiderivative of $\cos 3x$.

Solution The antiderivative of $\cos 3x$ is

$$\frac{1}{3}\sin 3x + C$$

since the derivative of this expression reproduces $\cos 3x$:

$$\frac{d}{dx}\left(\frac{1}{3}\sin 3x + C\right) = \frac{1}{3} \cdot 3\cos 3x + 0 = \cos 3x$$

Example 7
Find the antiderivative of

$$\frac{2x}{x^2 + 1}$$

Solution Since the numerator of this expression is the derivative of the denominator, we can employ the rule for the natural logarithm. Thus, the antiderivative of

$$\frac{2x}{x^2 + 1}$$

is

$$\ln(x^2 + 1) + C$$

Now that we have discussed how to find the antiderivates of several functions, we turn our attention to using it to help us find the integral of a function.

▶ Using the Antiderivative to Calculate the Integral of a Function

To develop our method of calculating the integral of a function, let's use the specific function, $f(x) = x^2$, and focus on the area under this curve between $x = 1$ and $x = 2$:

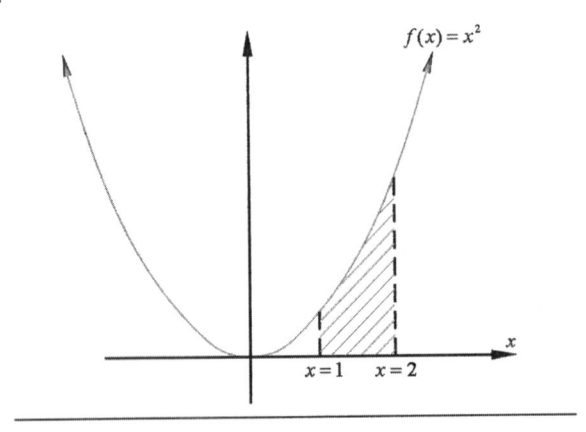

Fig. I.2.1

We can find the area by setting up the correct integral from our given information.

Beginning with our expression for the integral of a function

$$\text{Area} = \int_a^b f(x)\,dx$$

our area calculation becomes

$$\text{Area} = \int_1^2 x^2\,dx$$

Notice that we have substituted our two endpoints, 1 and 2, for a and b, as well as the function that is to be integrated, x^2, for $f(x)$.

We can now use antiderivatives to help us calculate the integral of this function. Since the integral of a function between two endpoints is also the area under the function between these two points, by calculating the integral we can find the required area.

The first step in solving the integral is to find the antiderivative of the function x^2. The antiderivative of x^2 is

$$\frac{1}{3}x^3 + C$$

We have found the antiderivative of the function, but we must still take the endpoints 1 and 2 into account. The mathematical expression that states we have found the antiderivative but have not yet dealt with the endpoints is

$$\int_1^2 x^2 dx = \left[\frac{1}{3}x^3 + C\right]_1^2$$

> **Math Review**
>
> Remember that the endpoints on the integration are often referred to as the *limits of integration.*

Notice that the two limits of integration, 1 and 2, have been placed on the brackets containing the antiderivative.

At this stage, you may be asking yourself what happened to the *dx* portion of the integral. This is where it is important to remember that in this section, we have moved directly to the shortcut method of finding integrals. Using this shortcut, it is sufficient for us simply to find the antiderivative of the function. If we instead used a more difficult, rigorous method of calculating our integral, the end result would still be the same.

Now we must deal with the endpoints that are attached to the integral. The endpoints are substituted for every x in the antiderivative. This substitution is a three-step process:

Substituting the Endpoints into the Antiderivative

1. Substitute the top endpoint into the antiderivative everywhere there is an x. The top endpoint is *always* substituted first, regardless of whether it is a positive number, a negative number, or a letter.
2. Substitute the bottom endpoint into the antiderivative everywhere there is an x. The bottom endpoint is *always* substituted second regardless of whether it is a positive number, a negative number, or a letter.
3. Subtract the antiderivative with the bottom endpoint substituted from the antiderivative with the top endpoint substituted.

Let's take these three steps to finish finding the area under $f(x) = x^2$ between $x = 1$ and $x = 2$. Substituting our top and bottom endpoints into the antiderivative gives us

$$\int_1^2 x^2 dx = \left[\frac{1}{3}x^3 + C\right]_1^2 = \left[\frac{1}{3}(2)^3 + C\right] - \left[\frac{1}{3}(1)^3 + C\right]$$

Notice that the first bracket in the subtraction contains the top endpoint substituted for x and the second bracket contains the bottom endpoint substituted for x. Also, note the minus sign between these two brackets. This minus sign corresponds to the third step in our substitution process.

Finally, we finish the arithmetic:

$$\int_1^2 x^2 dx = \left[\frac{1}{3}x^3 + C\right]_1^2 = \left[\frac{1}{3}(2)^3 + C\right] - \left[\frac{1}{3}(1)^3 + C\right]$$

$$\int_1^2 x^2 dx = \left[\frac{1}{3}(8) + C\right] - \left[\frac{1}{3}(1) + C\right]$$

$$\int_1^2 x^2 dx = \left[\frac{8}{3} + C\right] - \left[\frac{1}{3} + C\right]$$

$$\int_1^2 x^2 dx = \frac{8}{3} + C - \frac{1}{3} - C$$

$$\int_1^2 x^2 dx = \frac{7}{3}$$

The area under the function $f(x) = x^2$ between $x = 1$ and $x = 2$ is 7/3.

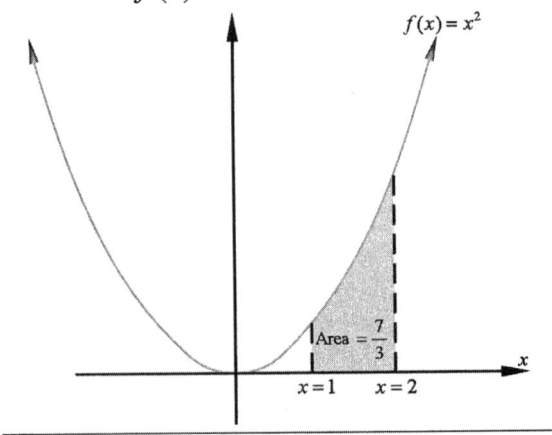

Fig. I.2.2

Notice that the unknown constant, C, cancelled once we substituted the endpoints. This will always be the case when we know the limits of integration. Because of this eventual cancellation, the unknown constant is often left out of the calculation. This cancellation is logical, since we are trying to find an actual area in cm^2, in^2, etc. We do not expect to have an unknown constant in our final area expression. The units for the area would depend on the units that we had attached to our axes when graphing the function. For example, if our axes had been constructed using inches, our area would be measured in square inches.

Finding the Area Under a Function

1. Find the antiderivative of the function beside dx in the integral. Be sure to include the C at the end of the antiderivative to allow for a possible unknown constant.
2. Place the top and bottom limits of integration on the outside right of the brackets containing the antiderivative.
3. To find a numerical solution, substitute the endpoints into the antiderivative in the correct order. Substitute the top endpoint (the top limit of integration first, and substitute the bottom endpoint (the bottom limit of integration) second.
4. Subtract the antiderivative expression with the bottom endpoint from the antiderivative expression with the top endpoint.
5. Make sure to place brackets around the second expression, since we must subtract the entire expression to get a cancellation of the unknown constant, C.

Example 5

Find the value of the integral of the function

$$f(x) = x^3$$

between $x = 0$ and $x = 1$.

Solution First, we write out the integral that we are trying to solve:

$$\int_0^1 x^3 dx$$

132   Integral Calculus

> **Math Review**
>
> To find the antiderivative to the right, we raise the exponent of 3 to 4, and invert the 4, getting ¼ as a coefficient in front of the antiderivative.

Next, we find the antiderivative of the function, and attach the limits of integration:

$$\int_0^1 x^3 dx = \left[\frac{1}{4}x^4 + C\right]_0^1$$

Finally, we substitute the endpoints and finish the arithmetic:

$$\int_0^1 x^3 dx = \left[\frac{1}{4}x^4 + C\right]_0^1$$

$$\int_0^1 x^3 dx = \left[\frac{1}{4}(1)^4 + C\right] - \left[\frac{1}{4}(0)^4 + C\right]$$

$$\int_0^1 x^3 dx = \left[\frac{1}{4}(1) + C\right] - [0 + C]$$

$$\int_0^1 x^3 dx = \frac{1}{4} + C - C$$

$$\int_0^1 x^3 dx = \frac{1}{4}$$

The area under the curve x^3 between $x = 0$ and $x = 1$ is ¼.

Example 6
Integrate

$$f(x) = 2x^2 + 5x$$

between $x = 1$ and $x = 3$.

Solution   First, we set up our integral:

$$\int_1^3 (2x^2 + 5x) dx$$

> **FYI**
>
> Notice that we only need one unknown constant, C, in the integral to the right even though we have two terms in the antiderivative.

In this example, finding the antiderivative of our function is slightly more difficult because it has two pieces. To find the antiderivative, we simply find the antiderivative of each term individually:

$$\int_1^3 (2x^2 + 5x)dx = \left[\frac{1}{3}(2)x^3 + \frac{1}{2}(5)x^2 + C\right]_1^3$$

Or, simplifying our antiderivative, we get

$$\int_1^3 (2x^2 + 5x)dx = \left[\frac{2}{3}x^3 + \frac{5}{2}x^2 + C\right]_1^3$$

Substituting the endpoints gives us

$$\int_1^3 (2x^2 + 5x)dx = \left[\frac{2}{3}x^3 + \frac{5}{2}x^2 + C\right]_1^3$$

$$\int_1^3 (2x^2 + 5x)dx = \left[\frac{2}{3}(3)^3 + \frac{5}{2}(3)^2 + C\right] - \left[\frac{2}{3}(1)^3 + \frac{5}{2}(1)^2 + C\right]$$

$$\int_1^3 (2x^2 + 5x)dx = [18 + 22.5 + C] - [0.67 + 2.5 + C]$$

$$\int_1^3 (2x^2 + 5x)dx = 37.33$$

The area under the function $2x^2 + 5x$ between $x = 1$ and $x = 3$ is 37.33.

▶ Definite and Indefinite Integrals

In this section, we will characterize our integrals. It is possible, and sometimes advantageous, to calculate the integral of a function without knowing the limits of integration. We can classify integrals according to whether or not we know the limits of integration.

> ### *Classifying Integrals*
>
> If we know the limits of integration (the endpoints), we call the integral a *definite* integral. If we do not know the limits of integration, we call the integral an *indefinite* integral.

Conveniently, we approach both definite and indefinite integrals in the same way. There is only one small difference between the methods of calculating the two different types of integrals:

1. To calculate a *definite* integral, first find the antiderivative of the function. Next, insert the endpoints. Remember that the unknown constant, C, will cancel when the limits of integration are inserted. The final answer will be a number.
2. To calculate an *indefinite* integral, simply find the antiderivative of the function and stop at this point. Because we do not know the endpoints, we will not be able to insert them and have the unknown constant, C, cancel. The unknown constant always remains as part of the final answer when calculating an integral of this type.

Example 7

Calculate the integral

$$\int (3x+4)dx$$

Solution First, we note that we do not know the limits of integration for this example. Therefore, it is an indefinite integral. This means that in order to solve the problem, all we need to do is find the antiderivative of the function:

$$\int (3x+4)dx = \frac{3}{2}x^2 + 4x + C$$

Because we have no endpoints to substitute, this is as far as we can go. Notice that the unknown constant, C, remains in our final answer.

Antiderivative Exercises with Solutions

Section 1

Find the antiderivative of the following functions. Remember to attach the unknown constant, C, to your final answer.

1. $f(x) = 2x$
2. $f(x) = 3x^2$

3. $f(x) = 6x + 4$

4. $f(x) = 5x^2 + 7x - 2$

5. $f(x) = 9x^3 - 3x$

Section 2

6. $f(x) = \sqrt{x}$

7. $f(x) = \dfrac{3}{x^2}$

8. $f(x) = 8x^2 - \sqrt[3]{x}$

9. $f(x) = \dfrac{4}{\sqrt{x}}$

10. $f(x) = x^5 - 3x^4 + \sqrt{x}$

Section 3

11. $f(x) = \cos 2x$

12. $f(x) = e^{3x}$

13. $f(x) = \dfrac{3x^2 + 2x}{x^3 + x^2 + 7}$

14. $f(x) = \sin 5x$

15. $f(x) = e^x + 6 \sin x$

Practice Antiderivative Exercises

Find the antiderivative of each of the following functions.

1. $f(x) = x^3$

2. $f(x) = 5x$

3. $f(x) = e^{7x}$

4. $f(x) = \dfrac{7x^6}{x^7 + 8}$

5. $f(x) = \cos 3x$

6. $f(x) = 2\sin 5x$

7. $f(x) = x^2 + 5x - 8$

8. $f(x) = \sqrt[4]{x}$

9. $f(x) = \dfrac{1}{x^3}$

10. $f(x) = e^x + 10$

Indefinite Integral Exercises with Solutions

Integrate the following indefinite integrals. Remember to include the constant of integration, C, in the final answer.

1. $\displaystyle\int x^2 \, dx$

2. $\displaystyle\int (x-5) \, dx$

3. $\int (4x^3 - 5x^2 + 7x)dx$

4. $\int \frac{1}{3}x^4 dx$

5. $\int \sqrt{x}\, dx$

6. $\int \sqrt[4]{x^3}\, dx$

7. $\int (6x^3 + 4x - 9)dx$

8. $\int \left(\frac{x}{4}\right)dx$

9. $\int (5x + \sqrt{x})dx$

10. $\int 4dx$

11. $\int \sin 2x\, dx$

12. $\int e^{8x} dx$

13. $\int (x^5 + \cos x)dx$

14. $\int \frac{1}{x} dx$

15. $\int (6 - e^x)dx$

Definite Integral Exercises with Solutions

Integrate the following definite integrals:

1. $\displaystyle\int_0^1 x^2\,dx$

2. $\displaystyle\int_1^3 (x^2-2)\,dx$

3. $\displaystyle\int_1^5 \frac{1}{2}x\,dx$

4. $\displaystyle\int_1^2 (3x^2-x)\,dx$

5. $\displaystyle\int_0^2 \sqrt{x}\,dx$

6. $\displaystyle\int_2^4 \frac{5}{x^2}\,dx$

7. $\displaystyle\int_{1.5}^{2.3} x^2\,dx$

8. $\displaystyle\int_0^{0.7} (3x-2)\,dx$

9. $\int_1^3 \dfrac{\sqrt[3]{x^2}}{4}\,dx$

10. $\int_2^3 (x+4)\,dx$

11. $\int_0^1 e^x\,dx$

12. $\int_0^{\pi/4} \cos 2x\,dx$

13. $\int_1^2 \dfrac{2x+3}{x^2+3x}\,dx$

14. $\int_1^5 dx$

15. $\int_0^{\pi/12} 12\sin 6x\,dx$

Practice Definite and Indefinite Integral Exercises

Calculate the following definite and indefinite integrals:

1. $\int x^3\,dx$

2. $\int_1^2 (x+2)\,dx$

140 Integral Calculus

3. $\displaystyle\int_0^3 x\,dx$

4. $\displaystyle\int \sqrt{x^3}\,dx$

5. $\displaystyle\int \frac{5}{x^2}\,dx$

6. $\displaystyle\int_1^2 (x^3 - x)\,dx$

7. $\displaystyle\int \left(\sqrt[3]{x} + x^5\right)dx$

8. $\displaystyle\int_{1.5}^{2.7} (4x+3)\,dx$

9. $\displaystyle\int \frac{1}{4}\,dx$

10. $\displaystyle\int_1^6 \frac{x}{8}\,dx$

11. $\displaystyle\int \sin 9x\,dx$

12. $\displaystyle\int_0^{\frac{\pi}{2}} \cos 3x\,dx$

13. $\displaystyle\int \frac{4x}{x^2+3}\,dx$

14. $\displaystyle\int 7\,dx$

15. $\displaystyle\int e^{6x}\,dx$

16. $\displaystyle\int y^2\,dy$

17. $\displaystyle\int_{\pi/2}^{\pi} \cos t\,dt$

18. $\displaystyle\int (\sin 3x + \cos 2x)\,dx$

19. $\displaystyle\int \frac{1}{t}\,dt$

20. $\displaystyle\int_0^1 (e^{2v} + v^2)\,dv$

INTEGRAL CALCULUS

3

The Method of Substitution

In the previous chapters on integration, we were able to integrate each of the functions by simply finding the antiderivative of each of the terms in the integral. In this section, we will discuss a method of integration known as the *Method of Substitution* that we can use for more difficult functions. This method can be applied to integrals that contain a large variety of functions, including polynomials, trigonometric functions, and many others. As we begin this section, it is important to note that although we are using a different method to find the integral of the function, we are still simply finding the area under the function between two endpoints.

▸ The Method of Substitution

We can us the method of substitution on functions similar to the following:

$$\int_1^2 (4x^2 + 3x)^5 (8x + 3) dx$$

$$\int (7x^3 - 5x^2)^3 (21x^2 - 10x) dx$$

$$\int_3^7 (8x^2 + 6x + 3)^5 (8x + 3) dx$$

$$\int (4x^4 - 12x^2)^7 (4x^3 - 6x) dx$$

Notice that in the first two examples, we have a term in parentheses that has been raised to a power and the exact derivative of the expression in the parentheses beside it. In the third example, we again have a term in parentheses raised to a power, but in this case the term next to it is one-half of the derivative of the term raised to a power. Lastly, in the fourth example we have a term in parentheses raised to a power and one-fourth of the derivative of the term beside it.

> We can use the *Method of Substitution* when we have an expression raised to a power and something *close* to the derivative of the expression beside it. The term beside the expression raised to a power must be either the derivative of the expression or be off by a simple numerical factor.

Our strategy to apply this method involves rewriting the integral in terms of the variable u. This rewrite will put the integral in a form that will allow us to easily find the antiderivative and solve the integral.

Example 1

To develop this method, we will use the specific function

$$\int (4x^2 + 5x)^6 (8x + 5) dx$$

First, we set the variable u equal to the term in parentheses that has been raised to the power:

$$u = 4x^2 + 5x$$

Next, we find the derivative of u with respect to the variable x:

$$\frac{du}{dx} = 8x + 5 \qquad (1)$$

Because the expression $\frac{du}{dx}$ is not just a notation for the derivative, but also represents the ratio of an infinitely small increment of u to an infinitely small increment of x, we can multiply both sides of Equation (1) by dx:

$$du = (8x + 5) dx$$

With this multiplication, notice that du is now the same as the second portion of our example integral. We can now use these two pieces, u and du, to rewrite our original integral:

$$\int \underbrace{(4x^2 + 5x)^6}_{u^6} \underbrace{(8x+5)dx}_{du} = \int u^6 du$$

Now that the integral is written in terms of the variable u, it is a simple matter to find its antiderivative:

$$\int u^6 \, du = \frac{1}{7} u^7 + C$$

The last step is to reinsert the x-dependent expression that u was representing.

Since $u = 4x^2 + 5x$,

$$\frac{1}{7} u^7 + C = \frac{1}{7} (4x^2 + 5x)^7 + C$$

Thus,

$$\int (4x^2 + 5x)^6 (8x + 5) \, dx = \frac{1}{7} (4x^2 + 5x)^7 + C$$

Using the method of substitution, we were able to solve the integral without having to deal with the original function. By substituting the variable u, the integral became much simpler and easier to solve.

In this next example, we use the method of substitution on an integral in which the term beside the expression raised to a power is not the exact derivative of the expression.

Example 2

$$\int (6x^2 + 8x)^5 (6x + 4) \, dx$$

First, we set the variable u equal to the term that has been raised to the power:

$$u = 6x^2 + 8x$$

Next, we find the derivative of u with respect to x:

$$\frac{du}{dx} = 12x + 8$$

Multiplying both sides by dx, we get

$$du = (12x + 8) \, dx$$

Notice that in this example, finding the derivative of *u* with respect to *x* does not yield the exact term in the original integral. Our *du* is off by a factor of 2:

$$(12x+8)dx = 2(6x+4)dx$$

Since the integral of a function is an actual area under a curve, we cannot simply multiply our original integral by a factor of 2 as this would double the area.

We can, however, multiply by a factor of 2 and a factor of ½ *at the same time*. Since multiplying by both 2 and ½ is the same as multiplying by 1, this will not change the value of the integral.

Thus, our strategy will be to multiply inside the integral by the factor of 2 necessary for the method of substitution and to multiply in front of the integral by a factor of ½ to compensate:

$$\int (6x^2+8x)^5 (6x+4)dx = \frac{1}{2}\int (6x^2+8x)^5 2(6x+4)dx$$

Distributing the 2 yields

$$\frac{1}{2}\int (6x^2+8x)^5 2(6x+4)dx = \frac{1}{2}\int (6x^2+5x)^5 (12x+8)dx$$

Notice that in this form, we can now use the method of substitution and rewrite the integral in terms of the variable *u*:

$$\frac{1}{2}\int \underbrace{(6x^2+5x)^5}_{u^5} \underbrace{(12x+8)dx}_{du} = \frac{1}{2}\int u^5 du$$

Finding the antiderivative in terms of the variable *u* gives us

$$\frac{1}{2}\int u^5 du = \frac{1}{2}\left(\frac{1}{6}u^6 + C\right) = \frac{1}{12}u^6 + C$$

Notice that although we distributed the factor of ½, we still express the unknown constant as *C* since one-half of an unknown constant is still an unknown constant.

Finally, reinserting the x-dependent expression that u was representing gives

$$\frac{1}{12}u^6 + C = \frac{1}{12}(6x^2 + 8x)^6 + C$$

Thus,

$$\int (6x^2 + 8x)^5 (6x + 4) dx = \frac{1}{12}(6x^2 + 8x)^6 + C$$

Example 3

In this last example, we apply the method of substitution to calculate a definite integral:

$$\int_0^1 (5x^2 + x)^3 (10x + 1) dx$$

We approach this problem in the same way as the two previous examples. The only difference will be the insertion of the limits of integration at the end of the problem.

First, we set u equal to the term that has been raised to the power:

$$u = 5x^2 + x$$

Calculating $\frac{du}{dx}$ and multiplying both sides by dx gives

$$du = (10x + 1) dx$$

Rewriting the integral in terms of the variable u yields

$$\int \underbrace{(5x^2 + x)^3}_{u^3} \underbrace{(10x + 1) dx}_{du} = \int u^3 du$$

Notice that we have not included the endpoints of the integration at this stage. Although we could redefine our limits of integration in terms of the variable u, it is not beneficial for us at this stage. For our purposes, it is simpler to insert the endpoints at the end of the process.

Finding the antiderivative in terms of the variable u yields

$$\int u^3 du = \frac{1}{4}u^4 + C$$

Reinserting the x-dependent expression that u was representing gives

$$\frac{1}{4}u^4 + C = \frac{1}{4}(5x^2 + x)^4 + C$$

Our definite integral thus takes the form:

$$\int_0^1 (5x^2 + x)^3 (10x + 1) dx = \left[\frac{1}{4}(5x^2 + x)^4 + C \right]_0^1$$

We finish the calculation by inserting the endpoints in exactly the same manner as discussed in the last section on definite integrals:

$$\int_0^1 (5x^2 + x)^3 (10x + 1) dx = \left[\frac{1}{4}(5x^2 + x)^4 + C \right]_0^1$$

$$\int_0^1 (5x^2 + x)^3 (10x + 1) dx = \left[\frac{1}{4}(5(1)^2 + 1)^4 + C \right] - \left[\frac{1}{4}(5(0)^2 + 0)^4 + C \right]$$

$$\int_0^1 (5x^2 + x)^3 (10x + 1) dx = 324$$

Exercises with Solutions

Use the method of substitution to evaluate the following definite and indefinite integrals:

1. $\int (3x^2 + 4)^3 6x \, dx$

2. $\int (7x^4 - 5x^2)^4 (28x^3 - 10x) dx$

The Method of Substitution

3. $\displaystyle\int (6x^3+5x+4)^4(18x^2+5)dx$

4. $\displaystyle\int_1^3 (2x^3+3x)^4(12x^2+6)dx$

5. $\displaystyle\int_0^1 (x^2+x)^4(2x+1)dx$

6. $\displaystyle\int (2x^7-8x^3)^5(14x^6-24x^2)dx$

7. $\displaystyle\int (4x^2+6x)^5(4x+3)dx$

8. $\displaystyle\int_2^3 6(6x-1)^4 dx$

9. $\displaystyle\int_0^2 (8x^2+4x+3)^3(4x+1)dx$

10. $\displaystyle\int \frac{(\sqrt{x}+3)^5}{2\sqrt{x}} dx$

Practice Exercises

Use the method of substitution to evaluate the following definite and indefinite integrals:

1. $\displaystyle\int (2x^2+3)4x\,dx$

2. $\displaystyle\int (x^4+8x+3)^7(8x^3+16)dx$

3. $\displaystyle\int (x^2+7x-2)^4(2x+7)dx$

4. $\displaystyle\int_1^3 (x+1)^5 dx$

5. $\displaystyle\int_0^1 (x^3+2x)^2(3x^2+2)dx$

6. $\displaystyle\int \sqrt{2x^2+3x}\,(4x+3)dx$

7. $\displaystyle\int (x^2+4x)^5(x+2)dx$

8. $\displaystyle\int \frac{2}{\sqrt[3]{x^2+6x}} \cdot (x+3)dx$

9. $\displaystyle\int_1^2 (x^3+3x^2)^2(x^2+2x)dx$

10. $\displaystyle\int_0^4 (x^2-1)^3 x\,dx$

INTEGRAL CALCULUS

4

Differentials II – The Relationship Between Integral and Differential Calculus

▶ Derivatives and Differentials

In Differentials I, we learned that we can express the derivative of a function such as

$$y = x^2$$

using differential notation:

$$\frac{dy}{dx} = 2x \qquad (1)$$

This notation was an alternative method of expressing the equation that would predict the slope of the tangent line drawn at a point on the function:

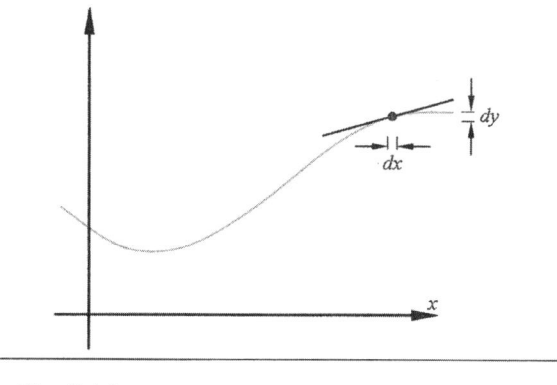

Fig. I.4.1

In this section, our purpose is to take the derivative expression in Equation (1) and use it to help us see the relationship between integral and differential calculus.

152   Integral Calculus

Our starting point is the realization that since the dx in the denominator of Equation (1) is an actual increment of length along the horizontal axis, we can multiply both sides of Equation (1) by this infinitesimal increment:

$$\frac{dy}{dx} = 2x$$

$$\frac{dy}{dx} \cdot dx = 2x\,dx$$

$$dy = 2x\,dx \qquad (2)$$

Equation (2) illustrates how an infinitesimal change along the y-axis relates to an infinitesimal change along the x-axis.

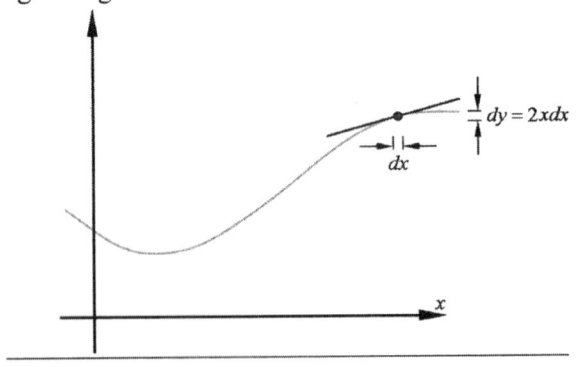

Fig. I.4.2

Notice that we can find the *total* change along the y-axis between points 1 and 2 by adding up all of the *infinitesimal* changes:

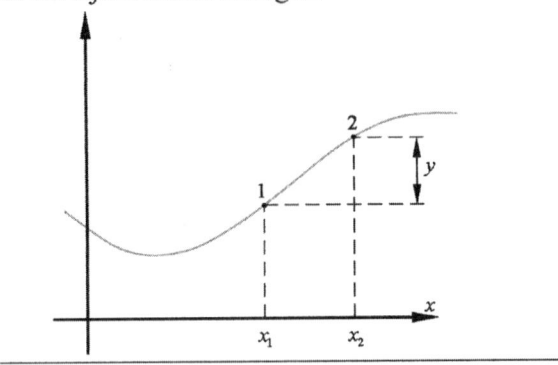

Fig. I.4.3

However, in order for us to execute this summation, we must integrate since we are adding together an infinite number of infinitely small increments.

Thus,

$$y = \int_{y_2}^{y_1} dy$$

Recalling the previous relationship between *dy* and *dx* given in Equation (2), to find the total change in *y* that results as we move from x_1 to x_2 along the function $y = x^2$, we calculate the integral:

$$y = \int_{x_1}^{x_2} 2x\, dx$$

Example 1

For the function $y = e^{5x}$, find the differential expression between an infinitely small change in the dependent variable, *y*, and the independent variable, *x*.

Solution First, we calculate the derivative of *y* with respect to *x* using the rule for the exponential function:

$$\frac{dy}{dx} = 5e^{5x}$$

Next, we multiply both sides of the equation by the increment *dx* to yield the solution:

$$\frac{dy}{dx} \cdot dx = 5e^{5x} dx$$
$$dy = 5e^{5x} dx$$

Example 2

Find the differential for the expression

$$y = \sin 3x$$

Solution First, we find the derivative of *y* with respect to *x*:

$$\frac{dy}{dx} = 3\cos 3x$$

Multiplying by *dx* yields

$$\frac{dy}{dx} \cdot dx = (3\cos 3x) dx$$
$$dy = (3\cos 3x) dx$$

Example 3

In electronics, the relationship between the current flow, i, and the electric charge, q is given by

$$\frac{dq}{dt} = i \quad (3)$$

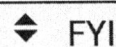

> **FYI**
>
> The SI standard unit for electric current is the *ampere* (amp).

since the current flow is the amount of electric charge that passes a point in a certain time interval. In the language of calculus, the current flow is the time rate of change of the electric charge.

Using this relationship between electric charge and current flow, find the total amount of electric charge that passes through a section of wire between $t = 0$ sec and $t = 1$ sec if the current flow in that section is given by

$$i = 3t^2$$

Solution We begin by substituting the current expression into Equation (3):

$$\frac{dq}{dt} = i$$

$$\frac{dq}{dt} = 3t^2 \quad (4)$$

Next, we multiply both sides of Equation (4) by the increment dt to find an expression for an infinitely small amount of charge, dq:

$$\frac{dq}{dt} \cdot dt = 3t^2 dt$$

$$dq = 3t^2 dt$$

To find the total amount of charge that passes during the time interval, we add together (or integrate) all of the infinitely small amounts of charge that pass through the section of the wire in that time period:

$$q = \int_{t_1}^{t_2} dq = \int_0^1 3t^2 dt$$

Executing this definite integral yields the total charge that passes between $t = 0$ sec and $t = 1$ sec:

$$q = \int_0^1 3t^2 dt = \left[t^3 + C\right]_0^1$$
$$q = \left[1^3 + C\right] - \left[0^3 + C\right]$$
$$q = 1$$

Thus, a total charge of 1 coulomb passed through the section of wire during the time interval.

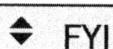

> **FYI**
>
> The SI standard unit for electric charge is the *coulomb* (C).

Example 4

Given that the rate of change of the dependent variable x with respect to the independent variable t for a function is

$$\frac{dx}{dt} = t^2 + 3 \qquad (5)$$

find the total change in the dependent variable x between $t = 1$ and $t = 2$.

Solution   First, we multiply both sides of Equation (5) by the increment dt to arrive at an expression for an infinitesimal change in the dependent variable x:

$$\frac{dx}{dt} \cdot dt = (t^2 + 3) dt$$
$$dx = (t^2 + 3) dt \qquad (6)$$

Next, we find the total change in the dependent variable, x, by integrating Equation (6) between $t = 1$ and $t = 2$:

$$x = \int_1^2 (t^2 + 3) dt = \left[\frac{1}{3}t^3 + 3t + C\right]_1^2$$
$$x = \left[\frac{1}{3}(2)^3 + 3(2) + C\right] - \left[\frac{1}{3}(1)^3 + 3(1) + C\right]$$
$$x = \frac{26}{3} - \frac{10}{3} = \frac{16}{3} = 5.33$$

Exercises with Solutions

Find the differential for each of the following functions:

1. $y = 5x^4$

2. $y = e^{3x}$

3. $B = \cos 5x$

4. $y = (2t^3 + 7t)^3$

5. $i = \tan t + \sqrt{t}$

Use differentials to find the change in the dependent variable as the independent variable is changed between the given limits.

6. $y = 2x^2$; $x = 0$ to $x = 1$

7. $y = e^x$; $x = 1$ to $x = 2$

8. $q = (t+1)^3$; $t = 0$ to $t = 3$

9. Given that the magnetic field B in a region changes with time according to the equation

$$\frac{dB}{dt} = e^{-2t}$$

use differentials to find the total change in the magnetic field from $t = 0$ sec to $t = 1$ sec.

10. The displacement, y, of a freefalling object is given by

$$y = 4.9t^2$$

Use differentials to find the displacement of the object between $t = 1$ sec and $t = 3$ sec.

Practice Exercises

Find the differential for each of the following functions:

1. $y = 6x^3$

2. $y = 7x^2 - 10x$

3. $T = \sin 8x$

4. $x = \sqrt[3]{t^2}$

5. $y = e^{3x} + \ln x$

Use differentials to find the change in the dependent variable as the independent variable is changed between the given limits.

6. $y = 7x^2;\quad x = 1$ to $x = 3$

7. $B = e^{5t};\quad t = 0$ to $t = 1$

8. $x = t^2 + 2t;\quad t = 2$ to $t = 3$

9. $v = \sin x;\quad x = 0$ to $x = \dfrac{\pi}{4}$

10. $y = \dfrac{1}{x^2};\quad x = 1$ to $x = 2$

11. The electric current through a certain circuit is given by

$$i = 2.4t^2$$

Find the total current flow through the circuit between $t = 2$ sec and $t = 4$ sec.

12. The momentum p of an object changes with time according to the equation

$$p = 2.3t + 4$$

Find the change in the momentum of the object between $t = 0$ sec and $t = 3$ sec.

13. The amount of a particular nuclear sample left after a time t is given by

$$N = e^{-100t}$$

Find the change in the amount of the sample between $t = 0.005$ sec and $t = 0.007$ sec.

14. For the following function, find the change in the dependent variable y if the independent variable x is varied between 1 and 5:

$$y = \frac{1}{x}$$

15. Find an equation that has a dependent variable of y and an independent variable of x that would generate the following differential expression:

$$dy = (3x^2 + 5x)dx$$

INTEGRAL CALCULUS

5

Integration by Parts

In Integral Calculus 3 - *The Method of Substitution*, we discussed how to integrate functions that were made of two pieces. As we discussed, in order to use that method, it was necessary to have a relationship between the two terms in the integral. One of the pieces had to be close to the derivative of the other piece. In this section, we move to an integration technique that is less restrictive. We can use this method, known as *Integration by Parts*, to integrate functions as diverse as

$$\int xe^{2x} dx$$

$$\int_0^4 x^2 \sin x \, dx$$

$$\int x \cos x \, dx$$

The advantage of this technique is that we can integrate functions that contain two pieces, and it is not necessary that one be the derivative of the other, as with the Method of Substitution.

▸ Integration by Parts

Integration by Parts is an algorithm that will produce the integral of the function. This algorithm is given by

$$\int u \, dv = uv - \int v \, du$$

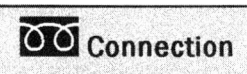

> **Connection**
>
> Notice that we are once again using the differential notation introduced in Differentials I and II.

To begin, we designate a portion of our integral as u. We then express the remaining portion of the integral (including dx) as dv. For example, if in the first integral listed above we choose u to be x,

$$u = x$$

then the remaining portion of the integral must be designated at dv:

$$dv = e^{2x} dx$$

In a similar manner, if we choose u to be x^2 in the second integral,

$$u = x^2$$

then dv is given by

$$dv = \sin x \, dx$$

Although we have complete freedom to choose which portion of the integral to call u and which to call dv, if we again look at the algorithm for Integration by Parts,

$$\int u \, dv = uv - \int v \, du$$

the right-hand side of the expression gives us guidance during the assigning process. Because the term du, the derivative of u, is necessary on the right-hand side, we assign u to the portion of the integral that will become simpler if its derivative is calculated. For example, in the first integral, if

$$u = x$$

then

$$\frac{du}{dx} = 1$$
$$du = dx$$

Or, in the second integral, if

$$u = x^2$$

then

$$\frac{du}{dx} = 2x$$
$$du = 2x\,dx$$

In both cases, when we calculated the derivative, it was simpler than the original expression. Notice also that in both examples, after finding the derivative of u with respect to x, we multiplied both sides of the equations by dx so that only du remained on the left-hand side.

Example 1

Use the technique of Integration by Parts to integrate the function

$$\int x e^{2x} dx$$

Solution First, we must choose a portion of the integral to designate as u. In this example, we have two choices which portion to assign u. We can either designate u as x, or u as e^{2x}. If we calculate the derivative of x and the derivative of e^{2x}, we see that the x term becomes simpler if its derivative is found.

$$u = x \qquad u = e^{2x}$$
$$\frac{du}{dx} = 1 \qquad \frac{du}{dx} = 2e^{2x}$$

Consequently, we choose u to be x. With this choice, the remainder of the integral, $e^{2x} dx$, is designated as dv:

$$dv = e^{2x} dx$$

Thus,

$$\int \underbrace{x}_{u} \underbrace{e^{2x} dx}_{dv}$$

On the right-hand side of the Integration by Parts algorithm, we must insert three items: u, v, and du:

$$\int u\,dv = uv - \int v\,du$$

Although we already have u, we must find v and du so that these terms can be substituted into the algorithm. To find du, we simply take the derivative of u:

$$u = x$$

$$\frac{du}{dx} = 1$$

and multiply both sides of the expression by dx:

$$\frac{du}{dx} \cdot dx = 1 \cdot dx$$

$$du = dx$$

Just as we used u to find du, we can use dv to find v. In this case, however, we find the *antiderivative* of dv rather than the derivative.

Beginning with dv and dividing both sides of the expression by dx gives us

$$dv = e^{2x} dx$$

$$\frac{dv}{dx} = \frac{e^{2x} dx}{dx}$$

$$\frac{dv}{dx} = e^{2x} \qquad (1)$$

Equation (1) tells us that the derivative of v with respect to x is equal to e^{2x}. Thus, if we find the antiderivative of Equation (1), we will have v.

Connection

For a full discussion of how to recover v from the differential element dv, see Differentials II.

If

$$\frac{dv}{dx} = e^{2x}$$

then

$$v = \frac{1}{2} e^{2x}$$

Note that we have not attached the usual unknown constant, C, to the antiderivative. This is not necessary since there will be an unknown constant on the final result from integrating by parts.

Now that we have u, du, and v, we can insert these results into the algorithm for Integration by Parts. The algorithm

$$\int u\,dv = uv - \int v\,du$$

becomes

$$\int \underbrace{x}_{u}\underbrace{e^{2x}dx}_{dv} = \underbrace{x}_{u}\underbrace{\left(\frac{1}{2}e^{2x}\right)}_{v} - \int \underbrace{\left(\frac{1}{2}e^{2x}\right)}_{v}\underbrace{dx}_{du} \quad (2)$$

Notice that the second term on the right-hand side of Equation (2) requires us to calculate another integral in order to finish integrating by parts. Completing this integral yields

$$\int xe^{2x}dx = x\left(\frac{1}{2}e^{2x}\right) - \frac{1}{4}e^{2x} + C$$

$$\int xe^{2x}dx = \frac{1}{2}xe^{2x} - \frac{1}{4}e^{2x} + C$$

Since the integral is an indefinite integral, we have the solution. If the integral were definite, we would now substitute the limits of integration into the antiderivative that the method of Integration by Parts has produced.

To summarize the method of Integration by Parts:

Integration by Parts

1. In the original integral, assign the letter u to the portion of the integral that will become simpler if its derivative is found.
2. The remaining portion of the integral that is not assigned u is designated as dv.
3. Calculate du by finding the derivative of u.
4. Calculate v by finding the antiderivative of dv.
5. Substitute u, du, and dv into the algorithm for Integration by Parts.
6. Calculate the integral on the right-hand side of the Integration by Parts algorithm.
7. If the integral is indefinite, attach the unknown constant, C.
8. If the integral is definite, substitute the endpoints and calculate the value of the integral.

Example 2

Use the method of Integration by Parts to calculate the integral

$$\int 2x \cos x \, dx$$

Solution First, we must decide which portion of the integral is going to be designated as *u*. Our two options are $2x$ and $\cos x$. Finding the derivative of each of these gives us

$$\frac{d}{dx}(2x) = 2 \qquad \frac{d}{dx}(\cos x) = -\sin x$$

Since $2x$ becomes simpler if its derivative is found, we select it as *u* and designate the remaining portion of the integral, $\cos x \, dx$, as *dv*.

$$\int \underbrace{2x}_{u} \underbrace{\cos x \, dx}_{dv}$$

Next, we find *du* by taking the derivative of *u*, and we find *v* by calculating the antiderivative of *dv*:

Derivative
$$\begin{array}{l} u = 2x \\ \dfrac{du}{dx} = 2 \\ du = 2dx \end{array} \qquad \begin{array}{l} dv = \cos x \, dx \\ \dfrac{dv}{dx} = \cos x \\ v = \sin x \end{array}$$ Antiderivative

Inserting *u*, *du*, and *v* into the algorithm for Integration by Parts,

$$\int u \, dv = uv - \int v \, du$$

yields

$$\int \underbrace{2x}_{u} \underbrace{\cos x \, dx}_{dv} = \underbrace{(2x)}_{u} \underbrace{\sin x}_{v} - \int \underbrace{\sin x}_{v} \underbrace{(2 \, dx)}_{du}$$

We can find the final integral on the right-hand side more readily if we move the factor of 2 outside the integral:

$$\int 2x\cos x\,dx = 2x\sin x - 2\int \sin x\,dx$$

Calculating the integral on the right-hand side yields the final result:

$$\int 2x\cos x\,dx = 2x\sin x - 2(-\cos x) + C$$

$$\int 2x\cos x\,dx = 2x\sin x + 2\cos x + C$$

As with Example 1, since the integral is indefinite, the antiderivative contains the unknown constant C.

Example 3
Use the method of Integration by Parts to evaluate the definite integral

$$\int_0^1 xe^{3x}\,dx$$

Solution We will approach this integral in the same way that we did Examples 1 and 2. After finding the antiderivative using the method of Integration by Parts, we will substitute the limits of integration and find a numerical value for the integral.

Selecting x to be u, and $e^{3x}dx$ to be dv, we find du and v:

Derivative
$u = x \qquad dv = e^{3x}dx$
$\dfrac{du}{dx} = 1 \qquad \dfrac{dv}{dx} = e^{3x}$ Antiderivative
$du = dx \qquad v = \dfrac{1}{3}e^{3x}$

Inserting u, du, and v into the algorithm for Integration by Parts,

$$\int u\,dv = uv - \int v\,du$$

yields

$$\int_0^1 \underbrace{x}_{u}\underbrace{e^{3x}dx}_{dv} = \left[\underbrace{x}_{u}\underbrace{\left(\frac{1}{3}e^{3x}\right)}_{v}\right]_0^1 - \int_0^1 \underbrace{\frac{1}{3}e^{3x}}_{v}\underbrace{dx}_{du}$$

Notice that we have attached the limits of integration, 0 and 1, to both terms on the right-hand side of the algorithm.

Completing the necessary integration on the second term of the right-hand side gives us

$$\int_0^1 xe^{3x}dx = \left[x\left(\frac{1}{3}e^{3x}\right)\right]_0^1 - \left[\frac{1}{9}e^{3x}\right]_0^1$$

Substituting the limits of integration yields

$$\int_0^1 xe^{3x}dx = \left[x\left(\frac{1}{3}e^{3x}\right)\right]_0^1 - \left[\frac{1}{9}e^{3x}\right]_0^1$$

$$\int_0^1 xe^{3x}dx = \left[1\left(\frac{1}{3}e^{3(1)}\right) - 0\left(\frac{1}{3}e^{3(0)}\right)\right] - \left[\frac{1}{9}e^{3(1)} - \frac{1}{9}e^{3(0)}\right]$$

$$\int_0^1 xe^{3x}dx = [6.695 - 0.333] - [2.232 - 0.111]$$

$$\int_0^1 xe^{3x}dx = [6.365] - [2.121] = 4.244$$

Example 4

Use the method of Integration by Parts to calculate the integral

$$\int x^2 e^{2x}dx$$

Solution In this example, we have the option to assign u to either x^2 or e^{2x}. If we find the derivative of each, we see that the x^2 term becomes simpler than the e^{2x} term:

$$\frac{d}{dx}(x^2) = 2x \qquad \frac{d}{dx}(e^{2x}) = 2e^{2x}$$

Consequently, we choose u to be x^2 and designate $e^{2x} dx$ as dv.

Calculating du and v yields

$$\text{Derivative} \begin{cases} u = x^2 \\ \dfrac{du}{dx} = 2x \\ du = 2x\,dx \end{cases} \qquad \begin{cases} dv = e^{2x} dx \\ \dfrac{dv}{dx} = e^{2x} \\ v = \dfrac{1}{2}e^{2x} \end{cases} \text{Antiderivative}$$

Inserting u, du, and v into the algorithm for Integration by Parts yields

$$\int \underbrace{x^2}_{u} \underbrace{e^{2x} dx}_{dv} = \underbrace{x^2}_{u}\underbrace{\left(\frac{1}{2}e^{2x}\right)}_{v} - \int \underbrace{\left(\frac{1}{2}e^{2x}\right)}_{v}\underbrace{(2x\,dx)}_{du}$$

Simplifying, we get

$$\int x^2 e^{2x} dx = \frac{1}{2}x^2 e^{2x} - \int xe^{2x} dx$$

At this point, we find ourselves in a slightly different position than in the previous examples. In Examples 1, 2, and 3, we were able to complete the required integral on the right-hand side of the Integration by Parts algorithm by calculating a simple antiderivative. Notice that in this example, we cannot find the required last integral so simply. To calculate the integral on the right-hand side, we must again use Integration by Parts.

Recalling that this integral is the same as that found in Example 1,

$$\int xe^{2x} dx = \frac{1}{2}xe^{2x} - \frac{1}{4}e^{2x} + C$$

we simply insert the result from that example:

$$\int x^2 e^{2x} dx = \frac{1}{2} x^2 e^{2x} - \int x e^{2x} dx$$

$$\int x^2 e^{2x} dx = \frac{1}{2} x^2 e^{2x} - \left[\frac{1}{2} x e^{2x} - \frac{1}{4} e^{2x} \right] + C$$

$$\int x^2 e^{2x} dx = \frac{1}{2} x^2 e^{2x} - \frac{1}{2} x e^{2x} + \frac{1}{4} e^{2x} + C$$

Notice that we have inserted one unknown constant to represent the possible unknown constants that arose during both applications of the Integration by Parts algorithm.

Exercises with Solutions

Use the method of Integration by Parts to calculate the following definite and indefinite integrals:

1. $\int x e^{4x} dx$

2. $\int x \sin x \, dx$

3. $\int_1^2 x e^x dx$

4. $\int t \cos 2t \, dt$

5. $\int x^2 \cos x \, dx$

Practice Exercises

Use the method of Integration by Parts to calculate the following definite and indefinite integrals:

1. $\int xe^{5x}\,dx$

2. $\int x\cos 6x\,dx$

3. $\int_{0}^{\frac{\pi}{2}} x\sin x\,dx$

4. $\int te^{7t}\,dt$

5. $\int x\sqrt{x+2}\,dx$

6. $\int_{-1}^{1} ye^{4y}\,dy$

7. $\int x^{2}\cos x\,dx$

8. $\int x\ln x$

9. $\int \sqrt{x}\ln x\,dx$

10. $\int \sin^{2} x\,dx$

INTEGRAL CALCULUS

6

Numerical Integration Techniques

> **♦ FYI**
>
> An *elementary* function is one that can be expressed as a combination of polynomial, exponential, logarithmic, and trigonometric functions.

Up until this point in our discussion of integration, the functions that we have integrated have all possessed antiderivatives that were simple, *elementary* functions. In many physical applications, however, the functions associated with the system do not possess such simple antiderivatives. This means that we cannot apply our integration technique of finding the antiderivative of the function and then inserting the endpoints. In this chapter, we will discuss numerical approximation techniques that can be used to integrate these more difficult functions.

▸ Endpoint Approximations

> **♦ FYI**
>
> If the area under the function between a and b is divided into n identical rectangles, the width of each rectangle can be represented as
>
> $$\Delta x = \frac{b-a}{n}$$

In the first of our approximation techniques, we use rectangles of identical widths to approximate the area under the function between two endpoints. These rectangles can be inserted so that their left side touches the function

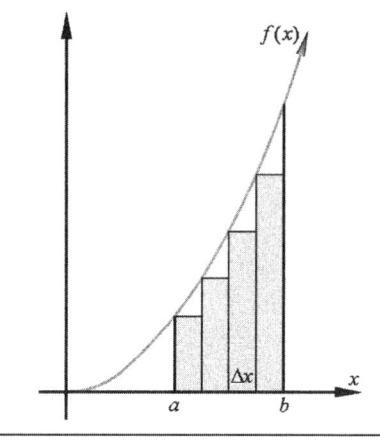

Fig. I.6.1

or so that their right side touches the function.

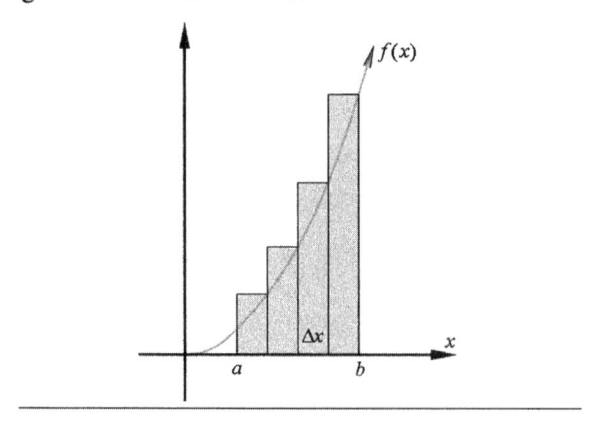

Fig. I.6.2

If we justify the left edge of the rectangle to the function, we call the approximation technique the *Left-endpoint Approximation*. If we justify the right edge of the rectangle to the function, we call the approximation technique the *Right-endpoint Approximation*.

▸ The Left-endpoint Approximation

Using the left-endpoint approximation and dividing the area under the function between a and b into n rectangles, we see that the height of each rectangle is determined by the x value associated with the left edge of the rectangle:

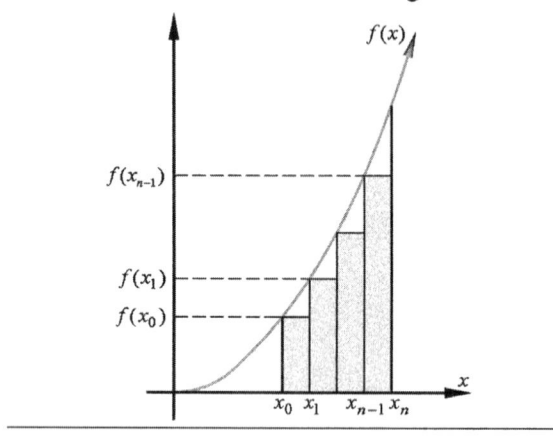

Fig. I.6.3

This means that an approximate area can be found by adding the individual areas of these rectangles,

$$\text{Area} = f(x_0)\Delta x + f(x_1)\Delta x + \ldots + f(x_{n-1})\Delta x$$

or, factoring off Δx,

$$\text{Area} = \Delta x [f(x_0) + f(x_1) + ... + f(x_{n-1})]$$

Notice that for the left-endpoint approximation the summation begins with the left end of the interval, x_0, and ends with x_{n-1}.

Example 1

Let's apply the left-endpoint approximation to an integral whose solution we already know, namely

$$\int_1^2 x^2 dx = \frac{7}{3}$$

$$\int_1^2 x^2 dx = 2.3333$$

The approximation to the area under the function will depend upon the number of rectangles that we use to divide the area between $x = 1$ and $x = 2$.

If we divide the area into 4 rectangles, the width of each rectangle becomes

$$\Delta x = \frac{b-a}{n}$$

$$\Delta x = \frac{2-1}{4}$$

$$\Delta x = 0.25$$

Thus, the left ends of the rectangles will be located at $x = 1$, 1.25, 1.5, and 1.75.

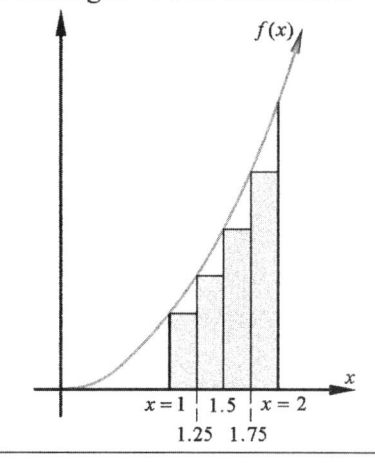

Fig. I.6.4

To find the height of each rectangle, we insert these x values into the function $f(x) = x^2$:

Rectangle	Location of Left End	Height
1	1.00	1.0000
2	1.25	1.5625
3	1.50	2.2500
4	1.75	3.0625

Since each rectangle must have the same width of 0.25, we can find the area of each rectangle by multiplying each rectangle's height by this width:

Rectangle	Height	Area
1	1.0000	0.2500
2	1.5625	0.3906
3	2.2500	0.5625
4	3.0625	0.7656

To find the approximation to the area under the curve, we simply add the areas of our four rectangles:

$$\text{Area} = 0.2500 + 0.3906 + 0.5625 + 0.7656$$
$$\text{Area} = 1.9687$$

Obviously, there is a discrepancy between this value and the actual area under the function, 2.3333. This discrepancy arises from the area that lies under the curve but was not included in any of our rectangles.

FYI

Note that regardless of the number of rectangles used, we use the left endpoint of the interval, a, to calculate the height of the first rectangle in the summation. However, the last rectangle in the summation uses the value of x that immediately precedes the right endpoint, b.

We can increase the accuracy of the approximation by increasing the number of rectangles dividing the area. As we decrease the width of our rectangles, as well as increase the number of rectangles, we will miss less and less area:

Number of rectangles, (n)	Width of each rectangle, (Δx)	Sum of the areas included in the rectangles
4	0.2500	1.9687
8	0.1250	2.1751
16	0.0625	2.2403

The larger the number of rectangles that we use to divide the area, the closer our approximation gets to the true value of 2.3333.

▸ The Right-endpoint Approximation

This approximation technique is almost identical to the left-endpoint approximation. In this case, however, we use the value of x associated with the right end of each rectangle to determine each rectangle's height.

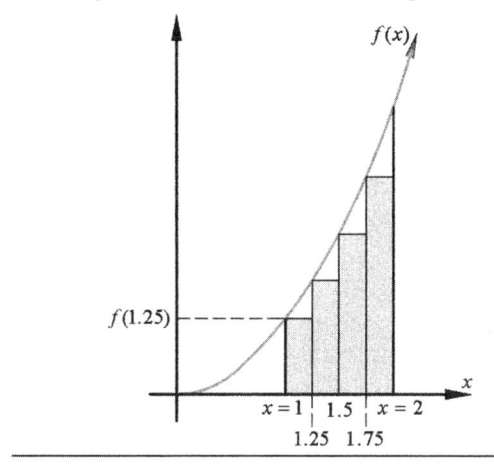

Fig. I.6.5

So, for example, where we used $x = 1$ to determine the height of the first rectangle in the summation using the left-endpoint approximation, we will use 1.25 to find this rectangle's height in the right-endpoint approximation.

As with the left-endpoint approximation, the approximation to the area under the curve becomes better as we increase the number of rectangles dividing the area between $x = 1$ and $x = 2$:

Number of rectangles, (n)	Width of each rectangle, (Δx)	Sum of the areas included in the rectangles
4	0.2500	2.7187
8	0.1250	2.5501
16	0.0625	2.4278

> **◆ FYI**
>
> Note that regardless of the number of rectangles used, we use the right-endpoint of the interval, b, to calculate the height of the last rectangle in the summation. However, the first rectangle uses the value of x that is immediately to the right of the left-endpoint of the interval, a.

Notice that in the case of the right-endpoint approximation, the discrepancy between our approximation and the true value is due to our rectangles including area that does not lie under the curve. As we use narrower rectangles, we decrease the amount of excess area included in the approximation.

The following table compares the left-endpoint approximation to the right-endpoint approximation for three different values of n:

Number of rectangles, (n)	Left-endpoint approximation	Right-endpoint approximation	Average of the two approximations
4	1.9687	2.7187	2.3437
8	2.1751	2.5501	2.3626
16	2.2403	2.4278	2.3340

Notice that the average of the two approximation techniques is closer to the true value of 2.3333, regardless of the number of rectangles. The fact that the average of the left-endpoint and right-endpoint approximations is closer to the true value is no coincidence. This average, which is known as the *Trapezoidal Approximation*, is our next approximation technique.

▸ The Trapezoidal Approximation

The average of the left-endpoint and right-endpoint approximations corresponds to justifying both the left and right ends of each rectangle on the function being analyzed. This means that the geometric form that we are using to divide the region of interest will not be a rectangle, but rather a *trapezoid*.

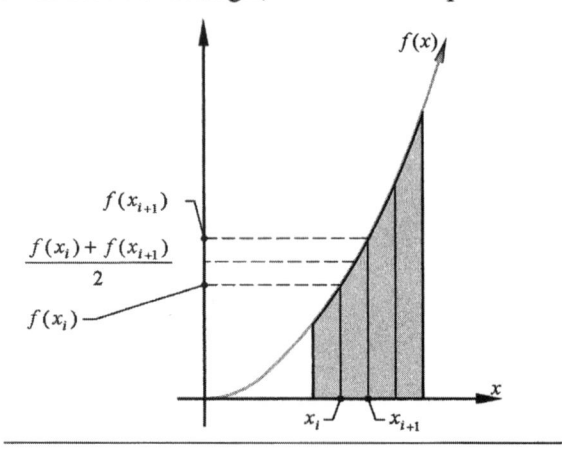

Fig. I.6.6

In order for us to arrive at a useful algorithm for finding the trapezoidal approximation to the area under a function, we must use an average of the expressions for both the left-endpoint and right-endpoint algorithms. Notice that because the height of each trapezoid is an average of the heights determined by the left and right endpoints, almost all of the points inside the interval will be used as the right end of one trapezoid and as the left endpoint of the trapezoid immediately adjacent. The only two points that will not be used in two trapezoid calculations are the left endpoint, a, and the right endpoint, b.

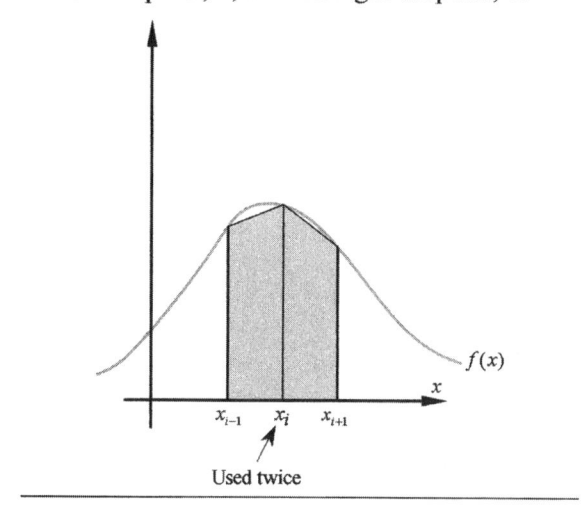

Fig. I.6.7

Thus, we can express the trapezoidal approximation to the area under the function as

$$\text{Area} = \left[\frac{f(x_0)+f(x_1)}{2}\right]\Delta x + \left[\frac{f(x_1)+f(x_2)}{2}\right]\Delta x + \ldots + \left[\frac{f(x_{n-1})+f(x_n)}{2}\right]\Delta x$$

or,

$$\text{Area} = \frac{\Delta x}{2}\left[f(x_0) + 2f(x_1) + \ldots 2f(x_{n-1}) + f(x_n)\right]$$

Notice that we multiply all factors in the area calculation by a factor of 2 except for the first term and the last term. This agrees with the fact that we use each point to find the average height of *two* trapezoids. The fact that we do not multiply $f(x_0)$ and $f(x_n)$ by a factor of 2 corresponds to our only using each of the endpoints of the interval to find the area of one trapezoid. The factor of 2 that divides Δx corresponds to the averaging process that is occurring.

Example 2

Use the trapezoidal approximation technique to evaluate the integral

$$\int_0^2 x^3 dx$$

using $n = 8$.

If the interval between $x = 0$ and $x = 2$ is divided into 8 subintervals, the width of each trapezoid becomes

$$\Delta x = \frac{b-a}{n}$$
$$\Delta x = \frac{2-0}{8}$$
$$\Delta x = 0.25$$

Thus, the trapezoidal approximation becomes

$$\text{Area} = \frac{0.25}{2}[f(0) + 2f(0.25) + 2f(0.5) + 2f(0.75) + \ldots + 2f(1.75) + f(2)]$$
$$\text{Area} = 0.125(32.5)$$
$$\text{Area} = 4.0625$$

If we compare our result with the true value for the integral

$$\int_0^2 x^3 dx = 4$$

we see that the trapezoidal approximation has produced a result remarkably close to the true result using only a very rough approximation ($n = 8$).

Example 3

In this example, we will see the true power of numerical integration. In the previous examples, the functions that were integrated possessed elementary antiderivatives. In direct contrast, it is well known that the function

$$f(x) = e^{-x^2}$$

does not have an elementary antiderivative. It is an example of an elementary function that does not possess an elementary antiderivative.

Problem:
Using the trapezoidal approximation, with $n = 10$, calculate the integral

$$\int_0^1 e^{-x^2} dx$$

> **FYI**
>
> The exponential function in this example is one of a family of functions known as *Gaussians*, named for the mathematician Karl Friedrich Gauss. This family of functions is extremely important in an enormous number of technological and scientific fields.

Using 10 subintervals, the width of each trapezoid becomes

$$\Delta x = \frac{b-a}{n}$$

$$\Delta x = \frac{1-0}{10}$$

$$\Delta x = 0.1$$

Thus, the trapezoidal approximation

$$\text{Area} = \frac{\Delta x}{2}\left[f(x_0) + 2f(x_1) + \ldots + 2f(x_{n-1}) + f(x_n)\right]$$

becomes

$$\text{Area} = \frac{0.1}{2}\left[f(0) + 2f(0.1) + 2f(0.2) + \ldots + 2f(0.9) + f(1)\right]$$

$$\text{Area} = 0.05(14.9261)$$

$$\text{Area} = 0.7463$$

Exercises with Solutions

Using the left-endpoint, right-endpoint, and trapezoidal approximation techniques, approximate the value of each of the following integrals:

1. $\int_0^1 2x^2 \, dx; \qquad n = 5$

2. $\int_1^2 e^x \, dx; \qquad n = 10$

3. $\int_1^2 (x^2 + 3) \, dx; \qquad n = 4$

180 Integral Calculus

4. $\int_{1}^{4} \dfrac{1}{x^3} dx$ $n = 8$

5. $\int_{1}^{4} \sqrt{x}\, dx;$ $n = 6$

Practice Exercises

Using the left-endpoint, right-endpoint, and trapezoidal approximation techniques, approximate the value of each of the following integrals:

1. $\int_{0}^{1} 3x^2 dx;$ $n = 5$

2. $\int_{1}^{2} \dfrac{1}{x^2} dx;$ $n = 8$

3. $\int_{0}^{1} e^{2x} dx;$ $n = 4$

4. $\int_{1}^{4} (x+2) dx;$ $n = 8$

5. $\int_{1}^{2} \sqrt[3]{x}\, dx;$ $n = 10$

INTEGRAL CALCULUS

Applications

1. If a force F is applied to a mass m, the work W done by the force can be found by multiplying the force by the distance that the mass moves:

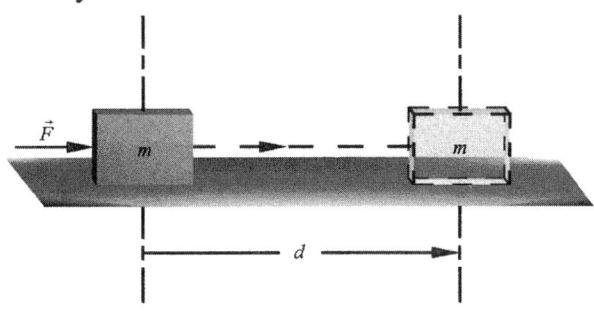

    $$W = Fd$$

    However, if the force exerted on the mass is not constant, but rather varies according to the location of the mass,

    $$F = F(x)$$

    we cannot find the amount of work done by the force so simply. To find the total work done as the variable force moves the mass from x_1 to x_2, we must integrate the force using these points as the limits of integration:

    $$W = \int_{x_1}^{x_2} F(x)\,dx$$

    Problem: Find the amount of work done by the force

    $$F(x) = 4x^2$$

    if the force moves a mass from $x = 1$ m to $x = 2$ m. Assume that the force is measured in newtons.

182   Integral Calculus

2. From Differential Calculus Application 5, we know that the current flow in a region relates to the electric charge in that region by

$$i = \frac{dq}{dt} \quad \textbf{(E1)}$$

Conversely, we can develop a method for finding the total charge that passes through a region by using the language of differentials. By multiplying both sides of Equation (E1) by the increment *dt*, we arrive at the relationship between an infinitesimal amount of charge and an infinitesimal amount of current:

$$i = \frac{dq}{dt}$$
$$idt = dq \quad \textbf{(E2)}$$

Thus, we can find the total amount of charge that passes by a certain point in a circuit during a time interval by integrating Equation (E2):

$$q = \int_{t_1}^{t_2} i(t)\, dt$$

Problem: The current flow in a section of a circuit is given by

$$i = 2t$$

Find the total charge, expressed in coulombs, that passes through the section between $t = 1$ sec and $t = 2$ sec.

3. The impulse *J* that a moving mass experiences can be found by multiplying the force *F* exerted on a mass by the time *t* that it is exerted.

$$J = Ft$$

However, if the force is not constant over time, but is instead time-dependent, we must integrate the force over the time interval to find the impulse caused by the force:

$$J = \int_{t_1}^{t_2} F(t)\, dt$$

Problem: Find the impulse caused by the force

$$F(t) = e^{2t}$$

between $t = 0$ sec and $t = 1$ sec.

4. In Differential Calculus Application 1, the relationship between the displacement s and the velocity v was given as

$$v = \frac{ds}{dt}$$

Drawing on differentials, this means that the displacement s that results from a time-varying velocity can be found using

$$s = \int_{t_1}^{t_2} v(t)\, dt$$

Problem: An object has a time-dependent velocity given by

$$v(t) = 3t^2 + 2t$$

   a) Find the displacement of the object between $t = 1$ sec and $t = 3$ sec. Assume the displacement is measured in meters.
   b) What effect, if any, did the unknown constant that accompanies the antiderivative have on the answer to part a)?

5. For a conservative system, the relationship between the potential energy stored in a system and the force exerted on the system is given by:

$$F = -\frac{d(PE)}{dx}$$

Using the material from Differentials II, construct an integral that would represent the total amount of potential energy stored due to a variable force $F(x)$ moving a mass between the points x_1 and x_2.

6. In a given interval $[a, b]$, a function $f(x)$ takes on a continuum of values. However, it is possible to find the *average value* that the function achieves in this interval by using:

$$f_{Avg} = \frac{1}{b-a} \int_a^b f(x)\, dx$$

Problem: Find the average value of each of the following functions in the given interval:

a) $f(x) = x^2$ $[0, 3]$

b) $f(x) = x + 5$ $[1, 2]$

c) $f(x) = \sin x$ $[0, 2\pi]$

7. In many technological situations, the *root-mean-square* (rms) value of a function is a more useful quantity than a simple average. For many quantities, such as voltage, power, etc., the rms value is the effective or useful amount of that quantity. The rms value of a function in an interval $[a, b]$ can be found by using:

$$f_{rms} = \sqrt{\frac{1}{b-a} \int_a^b [f(x)]^2 \, dx}$$

Notice that unlike the average value of a function (Integration Exercise 6), when finding the rms value, the function $f(x)$ is *squared* before we calculate the antiderivative of the function.

Problem: Find the rms value of each of the following functions in the given interval.

a) $f(x) = x^2$ $[0, 3]$

b) $f(x) = x + 2$ $[1, 2]$

c) $f(x) = 4$ $[1, 5]$

8. For an object such as a solid disc to rotate, a *torque* (τ) must be exerted on the object. If the torque causes the object to rotate through an angle θ, the torque has done *work* on the object.

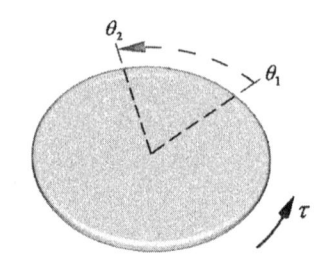

If the torque is constant as it rotates the object through an angle θ, we can find the work done by the torque using the relationship:

$$W = \tau\theta$$

However, if the torque is not constant, but rather varies depending upon the angle, we must integrate to find the work done by the torque as it rotates the object from θ_1 to θ_2:

$$W = \int_{\theta_1}^{\theta_2} \tau(\theta)\,d\theta$$

Problem: Find the work done by each of the following torques given the angle through which the torque rotates the object. Assume that all torques are measured in Nm, all angles are measured in radians, and the work done is measured in joules.

a) $\tau(\theta) = 3\theta^2$ $\quad\quad [0, \pi]$

b) $\tau(\theta) = 5\theta$ $\quad\quad \left[\dfrac{\pi}{4}, \dfrac{3\pi}{4}\right]$

c) $\tau(\theta) = 6$ $\quad\quad [0, 2\pi]$

9. The *moment of inertia* (I) of an object is a measurement of how easily the object is revolved around a particular axis. The moment of inertia of a point particle of mass m that is located a distance of r from the point of rotation is given by

$$I = mr^2$$

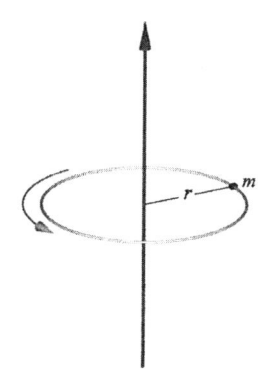

For a continuous object of mass *m*, we must draw upon calculus to find its moment of inertia since each increment of mass *dm* is located at a different distance from the point of rotation.

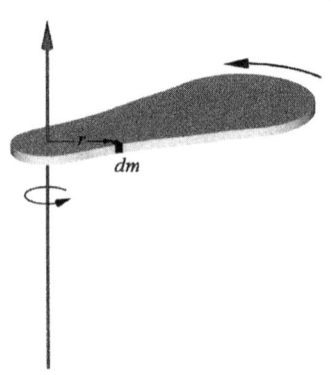

In the case of a continuous solid, we treat each increment of mass as a point mass and sum (integrate) each element's contribution to the total moment of inertia of the solid:

$$I = \int r^2 \, dm$$

Problem: Find an expression for the moment of inertia of a solid hoop of mass *m* and radius *r* if it is rotated around its central axis. *Hint:* Notice that each increment of mass is located at the same distance from the point of rotation, making *r* a constant in the integral.

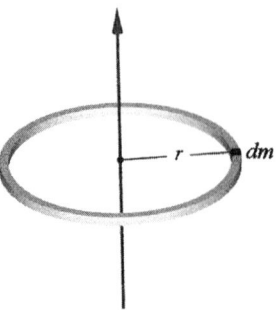

10. We can find the amount of heat *Q* required for a mass *m* to experience a temperature change of ΔT by using

$$Q = mc\Delta T$$

where the variable *c* represents the *specific heat capacity* of the material. In many cases, we assume that the specific heat capacity of a material is a constant. However, the specific heat of a material, extracted from a table of physical constants, is really an average of the specific heat capacity of the material over the temperature range ΔT.

Applications 187

In actuality, the specific heat capacity of a material varies continuously. Consequently, if we require more precision in the calculation, we must use the language of differentials. An infinitely small change in a material's temperature, dT, results from the addition of an infinitely small amount of heat, dQ:

$$dQ = mcdT$$

We can then use this differential relationship to construct an integral that will yield the amount of heat required to change the temperature of a material that has a temperature-dependent specific heat capacity:

$$Q = m \int_{T_i}^{T_f} cdT$$

Problem: A particular substance is found to have a specific heat capacity over a range of temperatures given by:

$$c(T) = 1 + 0.02T$$

Find the amount of heat required to change 500 g of the material from 293 K to 300 K.

Solutions

SOLUTIONS FOR DIFFERENTIAL CALCULUS 1:
Derivative Preliminaries

pp. 19 - 20

In the following equations, identify the dependent and independent variable:

1. $y = 5x^2 - 7x$
   Dependent – y, Independent – x

2. $B = 7t^3$
   Dependent – B, Independent – t

3. $R = 3w + 8$
   Dependent – R, Independent – w

4. $E = 9s - 11$
   Dependent – E, Independent – s

Given the following sets of points, calculate the slope of the line that would pass through the points:

5. For the points $(4, 2), (3, 1)$

    $\text{slope} = \dfrac{y_2 - y_1}{x_2 - x_1}$

    $\text{slope} = \dfrac{1-2}{3-4} = \dfrac{-1}{-1}$

    $\text{slope} = 1$

6. For the points $(2, 5), (8, 1)$

    $\text{slope} = \dfrac{y_2 - y_1}{x_2 - x_1}$

    $\text{slope} = \dfrac{1-5}{8-2} = \dfrac{-4}{6}$

    $\text{slope} = \dfrac{-2}{3}$

7. For the points $(3, 5), (9, 5)$

    $\text{slope} = \dfrac{y_2 - y_1}{x_2 - x_1}$

    $\text{slope} = \dfrac{5-5}{9-3} = \dfrac{0}{6}$

    $\text{slope} = 0$

8. For the points $(0, 0), (2, 6)$

    $\text{slope} = \dfrac{y_2 - y_1}{x_2 - x_1}$

    $\text{slope} = \dfrac{6-0}{2-0} = \dfrac{6}{2}$

    $\text{slope} = 3$

For each of the following functions, find $f(0)$ and $f(1)$:

9. $f(x) = 2x$

    $f(0) = 0, f(1) = 2$

10. $f(x) = 3x^2$

    $f(0) = 0, f(1) = 3$

11. $f(x) = 9 - 4x$

$f(0) = 9, f(1) = 5$

12. $f(x) = 4x^2 + 5$

$f(0) = 5, f(1) = 9$

Calculate the limit of the following expressions:

13. $\lim\limits_{x \to 2}(4x + 1) = 4(2) + 1 = 9$

14. $\lim\limits_{x \to 0}(3x^2 + 4x + 8) = 3(0)^2 + 4(0) + 8 = 8$

15. $\lim\limits_{x \to 2} \dfrac{x^2 - 4}{x - 2} = \lim\limits_{x \to 2} \dfrac{(x+2)(x-2)}{x-2} = \lim\limits_{x \to 2}(x + 2) = 2 + 2 = 4$

SOLUTIONS FOR DIFFERENTIAL CALCULUS 2:
The Derivative of a Function

p. 35

Use the definition of the derivative

$$f'(x) = \lim_{\Delta x \to 0} \frac{f(x+\Delta x)-f(x)}{\Delta x}$$

to calculate the derivatives of the following functions:

1. $f(x) = 2x$

$f'(x) = \lim_{\Delta x \to 0} \dfrac{f(x+\Delta x)-f(x)}{\Delta x}$

$f'(x) = \lim_{\Delta x \to 0} \dfrac{2(x+\Delta x)-2x}{\Delta x}$

$f'(x) = \lim_{\Delta x \to 0} \dfrac{2x+2\Delta x-2x}{\Delta x}$

$f'(x) = \lim_{\Delta x \to 0} \dfrac{2\Delta x}{\Delta x}$

$f'(x) = \lim_{\Delta x \to 0} 2$

$f'(x) = 2$

2. $f(x) = 3x^2$

$f'(x) = \lim_{\Delta x \to 0} \dfrac{f(x+\Delta x)-f(x)}{\Delta x}$

$f'(x) = \lim_{\Delta x \to 0} \dfrac{3(x+\Delta x)^2 - 3x^2}{\Delta x}$

$f'(x) = \lim_{\Delta x \to 0} \dfrac{3(x^2 + 2x\Delta x + (\Delta x)^2) - 3x^2}{\Delta x}$

$f'(x) = \lim_{\Delta x \to 0} \dfrac{3x^2 + 6x\Delta x + 3(\Delta x)^2 - 3x^2}{\Delta x}$

$f'(x) = \lim_{\Delta x \to 0} \dfrac{6x\Delta x + 3(\Delta x)^2}{\Delta x}$

$f'(x) = \lim_{\Delta x \to 0} \dfrac{\Delta x(6x + 3\Delta x)}{\Delta x}$

$f'(x) = \lim_{\Delta x \to 0} (6x + 3\Delta x)$

$f'(x) = 6x + 3(0)$

$f'(x) = 6x$

3. $f(x) = 5x - 3$

$$f'(x) = \lim_{\Delta x \to 0} \frac{f(x+\Delta x) - f(x)}{\Delta x}$$
$$f'(x) = \lim_{\Delta x \to 0} \frac{5(x+\Delta x) - 3 - (5x - 3)}{\Delta x}$$
$$f'(x) = \lim_{\Delta x \to 0} \frac{5x + 5\Delta x - 3 - 5x + 3}{\Delta x}$$
$$f'(x) = \lim_{\Delta x \to 0} \frac{5\Delta x}{\Delta x}$$
$$f'(x) = \lim_{\Delta x \to 0} 5$$
$$f'(x) = 5$$

4. $f(x) = 4x^3$

$$f'(x) = \lim_{\Delta x \to 0} \frac{f(x+\Delta x) - f(x)}{\Delta x}$$
$$f'(x) = \lim_{\Delta x \to 0} \frac{4(x+\Delta x)^3 - 4x^3}{\Delta x}$$
$$f'(x) = \lim_{\Delta x \to 0} \frac{4\left(x^3 + 3x^2\Delta x + 3x(\Delta x)^2 + (\Delta x)^3\right) - 4x^3}{\Delta x}$$
$$f'(x) = \lim_{\Delta x \to 0} \frac{4x^3 + 12x^2\Delta x + 12x(\Delta x)^2 + 4(\Delta x)^3 - 4x^3}{\Delta x}$$
$$f'(x) = \lim_{\Delta x \to 0} \frac{12x^2\Delta x + 12x(\Delta x)^2 + 4(\Delta x)^3}{\Delta x}$$
$$f'(x) = \lim_{\Delta x \to 0} \frac{\Delta x\left(12x^2 + 12x\Delta x + 4(\Delta x)^2\right)}{\Delta x}$$
$$f'(x) = \lim_{\Delta x \to 0} \left(12x^2 + 12x\Delta x + 4(\Delta x)^2\right)$$
$$f'(x) = 12x^2 + 12x(0) + 4(0)^2$$
$$f'(x) = 12x^2$$

5. $f(x) = 6x^2 + 7x$

$f'(x) = \lim_{\Delta x \to 0} \dfrac{f(x+\Delta x) - f(x)}{\Delta x}$

$f'(x) = \lim_{\Delta x \to 0} \dfrac{6(x+\Delta x)^2 + 7(x+\Delta x) - (6x^2 + 7x)}{\Delta x}$

$f'(x) = \lim_{\Delta x \to 0} \dfrac{6(x^2 + 2x\Delta x + (\Delta x)^2) + 7x + 7\Delta x - 6x^2 - 7x}{\Delta x}$

$f'(x) = \lim_{\Delta x \to 0} \dfrac{6x^2 + 12x\Delta x + 6(\Delta x)^2 + 7x + 7\Delta x - 6x^2 - 7x}{\Delta x}$

$f'(x) = \lim_{\Delta x \to 0} \dfrac{12x\Delta x + 6(\Delta x)^2 + 7\Delta x}{\Delta x}$

$f'(x) = \lim_{\Delta x \to 0} \dfrac{\Delta x(12x + 6\Delta x + 7)}{\Delta x}$

$f'(x) = \lim_{\Delta x \to 0} (12x + 6\Delta x + 7)$

$f'(x) = 12x + 6(0) + 7$

$f'(x) = 12x + 7$

SOLUTIONS FOR DIFFERENTIAL CALCULUS 3: Rules for Finding The Derivatives of Functions

pp. 43 - 45

Find the derivatives of the following functions using the rule for polynomials:

Section 1

1. $f(x) = 6x^2 + 4x + 3$

    $f'(x) = 12x + 4$

2. $f(x) = 5x - 4$

    $f'(x) = 5$

3. $f(x) = 7x^5 - 2x^4$

    $f'(x) = 35x^4 - 8x^3$

4. $f(x) = -8x^2 + 15$

    $f'(x) = -16x$

5. $f(x) = 2x - 5x^3$

    $f'(x) = 2 - 15x^2$

6. $f(x) = -4x + 19$

    $f'(x) = -4$

7. $f(x) = 14x^2 + 7x - 6.75$

    $f'(x) = 28x + 7$

8. $f(x) = 8.3x + 7$

    $f'(x) = 8.3$

9. $f(x) = 9$

    $f'(x) = 0$

10. $f(x) = 12x - 4.3x^2 + 5x^4$

    $f'(x) = 12 - 8.6x + 20x^3$

Section 2

11. $f(x) = \dfrac{3}{x^2}$

    $f(x) = 3x^{-2}$

    $f'(x) = -6x^{-3}$

12. $f(x) = \sqrt[3]{x^4}$

    $f(x) = x^{\frac{4}{3}}$

    $f'(x) = \dfrac{4}{3} x^{\frac{1}{3}}$

Solutions 197

13. $f(x) = \sqrt{x}$

$f(x) = x^{\frac{1}{2}}$

$f'(x) = \frac{1}{2}x^{-\frac{1}{2}}$

14. $f(x) = x^3 + \sqrt{x}$

$f(x) = x^3 + x^{\frac{1}{2}}$

$f'(x) = 3x^2 + \frac{1}{2}x^{-\frac{1}{2}}$

15. $f(x) = \frac{4}{\sqrt[3]{x}}$

$f(x) = 4x^{-\frac{1}{3}}$

$f'(x) = -\frac{4}{3}x^{-\frac{4}{3}}$

16. $f(x) = 4x^3 - 5x^2 + \frac{7}{x} + 3$

$f(x) = 4x^3 - 5x^2 + 7x^{-1} + 3$

$f'(x) = 12x^2 - 10x - 7x^{-2}$

17. $f(x) = \frac{x^2 - 4}{2}$

$f(x) = \frac{1}{2}x^2 - \frac{4}{2}$

$f(x) = \frac{1}{2}x^2 - 2$

$f'(x) = x$

18. $f(x) = \frac{4}{x^5} + \frac{3}{x^2} + 5$

$f(x) = 4x^{-5} + 3x^{-2} + 5$

$f'(x) = -20x^{-6} - 6x^{-3}$

19. $f(x) = \frac{3x^2 + 4x}{3}$

$f(x) = \frac{3x^2}{3} + \frac{4x}{3}$

$f(x) = x^2 + \frac{4}{3}x$

$f'(x) = 2x + \frac{4}{3}$

20. $f(x) = \frac{x^3}{\sqrt{x}}$

$f(x) = \frac{x^3}{x^{\frac{1}{2}}}$

$f(x) = x^{\frac{5}{2}}$

$f'(x) = \frac{5}{2}x^{\frac{3}{2}}$

Solutions

pp. 51 - 52

Find the derivative of each of the following functions:

Section 1

1. $f(x) = \sin(3x)$

   $f'(x) = 3\cos(3x)$

2. $f(x) = \ln(2x)$

   $f'(x) = \dfrac{2}{2x} = \dfrac{1}{x}$

3. $f(x) = \sin x$

   $f'(x) = \cos x$

4. $f(x) = 6\ln(7x^2 + 9x)$

   $f'(x) = 6 \cdot \dfrac{14x+9}{7x^2+9x} = \dfrac{84x+54}{7x^2+9x}$

5. $f(x) = \cos(3x^4)$

   $f'(x) = -12x^3 \sin(3x^4)$

6. $f(x) = -\ln x$

   $f'(x) = -\dfrac{1}{x}$

7. $f(x) = \cos x$

   $f'(x) = -\sin x$

8. $f(x) = -e^{3x}$

   $f'(x) = -3e^{3x}$

9. $f(x) = e^{7x+8}$

   $f'(x) = 7e^{7x+8}$

10. $f(x) = \cos(7x^3 - 5x^2 + 10x)$

    $f'(x) = -(21x^2 - 10x + 10)\sin(7x^3 - 5x^2 + 10x)$

11. $f(x) = e^x$

    $f'(x) = e^x$

12. $f(x) = 8\sin(x^2)$

    $f'(x) = 8(2x)\cos(x^2) = 16x\cos(x^2)$

Section 2 – Mixed Types

13. $f(x) = \sqrt{x}$

    $f(x) = x^{\frac{1}{2}}$

    $f'(x) = \frac{1}{2}x^{-\frac{1}{2}}$

14. $f(x) = \frac{8}{x^5} - 4x^3$

    $f(x) = 8x^{-5} - 4x^3$

    $f'(x) = -40x^{-6} - 12x^2$

15. $f(x) = 6\cos(9x^3 - 5x^2)$

    $f'(x) = -6(27x^2 - 10x)\sin(9x^3 - 5x^2)$

16. $f(x) = e^{7x^3 - 5x}$

    $f'(x) = (21x^2 - 5)e^{7x^3 - 5x}$

17. $f(x) = e^{\sqrt{x}}$

    $f(x) = e^{x^{\frac{1}{2}}}$

    $f'(x) = \frac{1}{2}x^{-\frac{1}{2}}e^{x^{\frac{1}{2}}}$

18. $f(x) = \cos(5x - 3) + \sqrt[3]{x^4}$

    $f(x) = \cos(5x - 3) + x^{\frac{4}{3}}$

    $f'(x) = -5\sin(5x - 3) + \frac{4}{3}x^{\frac{1}{3}}$

19. $f(x) = \frac{5}{x^6} + \sin(9x)$

    $f(x) = 5x^{-6} + \sin(9x)$

    $f'(x) = -30x^{-7} + 9\cos(9x)$

20. $f(x) = \cos(5x^2 - 12x) + \sin(6x - 1)$

    $f'(x) = -(10x - 12)\sin(5x^2 - 12x) + 6\cos(6x - 1)$

21. $f(x) = e^{7x - 3x^4} + \sin(8x) + 2x^5$

    $f'(x) = (7 - 12x^3)e^{7x - 3x^4} + 8\cos(8x) + 10x^4$

22. $f(x) = \ln(2x) + \sin x$

    $f'(x) = \frac{2}{2x} + \cos x$

    $f'(x) = \frac{1}{x} + \cos x$

23. $f(x) = \cos(2x^2) - \ln(6x - 1)$

    $f'(x) = -4x\sin(2x^2) - \frac{6}{6x - 1}$

24. $f(x) = 5e^{4.1x}$

$f'(x) = 5(4.1)e^{4.1x} = 20.5e^{4.1x}$

25. $f(x) = 3.2\ln(2x+1)$

$f'(x) = 3.2\left(\dfrac{2}{2x+1}\right) = \dfrac{6.4}{2x+1}$

pp. 62 - 65

Section 1

Use the Product Rule, Quotient Rule, or Power Rule to find the derivatives of these functions involving simple polynomials.

1. $f(x) = (3x^2 + 8x)(5x^3 + 7x^2 - 6x)$

   Using the Product Rule:

   $f'(x) = (3x^2 + 8x)(15x^2 + 14x - 6) + (5x^3 + 7x^2 - 6x)(6x + 8)$

2. $f(x) = (4x^5 + 5x^2 + 3x)(9x^2 - 6x)$

   Using the Product Rule:

   $f'(x) = (4x^5 + 5x^2 + 3x)(18x - 6) + (9x^2 - 6x)(20x^4 + 10x + 3)$

3. $f(x) = \dfrac{x^3 + 4x^2}{7x^4 + 3x^3}$

   Using the Quotient Rule:

   $f'(x) = \dfrac{(7x^4 + 3x^3)(3x^2 + 8x) - (x^3 + 4x^2)(28x^3 + 9x^2)}{(7x^4 + 3x^3)^2}$

4. $f(x) = \dfrac{9x - 1}{x^5 - 6x^3}$

   Using the Quotient Rule:

   $f'(x) = \dfrac{(x^5 - 6x^3)(9) - (9x - 1)(5x^4 - 18x^2)}{(x^5 - 6x^3)^2}$

Solutions   201

5. $f(x) = (4x^3 + 5x^2 - 6x)^4$

   Using the Power Rule:

   $f'(x) = 4(4x^3 + 5x^2 - 6x)^3 (12x^2 + 10x - 6)$

6. $f(x) = (7x^2 - 11x + 10)^6$

   Using the Power Rule:

   $f'(x) = 6(7x^2 - 11x + 10)^5 (14x - 11)$

7. $f(x) = (6x^3 + 5x)(4x^4 - 7x^2)$

   Using the Product Rule:

   $f'(x) = (6x^3 + 5x)(16x^3 - 14x) + (4x^4 - 7x^2)(18x^2 + 5)$

8. $f(x) = (5x^7 - 6x^4 + 9x^2)^5$

   Using the Power Rule:

   $f'(x) = 5(5x^7 - 6x^4 + 9x^2)^4 (35x^6 - 24x^3 + 18x)$

9. $f(x) = \dfrac{5x^4 + 3x^2 - 9x}{4x^2 - 7.5x + 3.2}$

   Using the Quotient Rule:

   $f'(x) = \dfrac{(4x^2 - 7.5x + 3.2)(20x^3 + 6x - 9) - (5x^4 + 3x^2 - 9x)(8x - 7.5)}{(4x^2 - 7.5x + 3.2)^2}$

202 Solutions

10. $f(x) = (5x^3 + 7x)^4 + \dfrac{3x^3 - 4x^2}{9x^2 - 8x + 2}$

Using the Power Rule on the first term and the Quotient Rule on the second term:

$f'(x) = 4(5x^3 + 7x)^3(15x^2 + 7) + \dfrac{(9x^2 - 8x + 2)(9x^2 - 8x) - (3x^3 - 4x^2)(18x - 8)}{(9x^2 - 8x + 2)^2}$

Section 2

Use the Product Rule, Quotient Rule, or Power Rule to find the derivatives of the following functions.

11. $f(x) = \dfrac{e^{4x^3 + 6x^2}}{3x^4}$

Using the Quotient Rule:

$f'(x) = \dfrac{3x^4(12x^2 + 12x)e^{4x^3 + 6x^2} - e^{4x^3 + 6x^2}(12x^3)}{(3x^4)^2}$

12. $f(t) = (7t^2 - 6t + 9)\sin(5t^3 + 8t)$

Using the Product Rule:

$f'(t) = (7t^2 - 6t + 9)(15t^2 + 8)\cos(5t^3 + 8t) + \sin(5t^3 + 8t)(14t - 6)$

13. $f(t) = (5t + e^{7t})^4$

Using the Power Rule:

$f'(t) = 4(5t + e^{7t})^3(5 + 7e^{7t})$

14. $f(x) = \dfrac{\sin(3x-2)}{\sqrt{x^5}}$

First rewriting the function and then using the Quotient Rule:

$$f(x) = \dfrac{\sin(3x-2)}{x^{\frac{5}{2}}}$$

$$f'(x) = \dfrac{x^{\frac{5}{2}}[3\cos(3x-2)] - \sin(3x-2)\left(\dfrac{5}{2}x^{\frac{3}{2}}\right)}{\left(x^{\frac{5}{2}}\right)^2}$$

15. $f(p) = (6p + \sqrt{p})(5p^3 + 8p^2 - 12p)$

First rewriting the radical and then using the Product Rule:

$$f(p) = \left(6p + p^{\frac{1}{2}}\right)(5p^3 + 8p^2 - 12p)$$

$$f'(p) = \left(6p + p^{\frac{1}{2}}\right)(15p^2 + 16p - 12) + (5p^3 + 8p^2 - 12p)\left(6 + \dfrac{1}{2}p^{-\frac{1}{2}}\right)$$

Summary Exercises

Find the derivative of the following functions.

1. $f(x) = 7x^3 - 5x$

$f'(x) = 21x^2 - 5$

2. $f(t) = (3t - 5)^7$

$f'(t) = 7(3t - 5)^6 (3)$

3. $f(t) = \sin(9t^3 - 4t)$

$f'(t) = (27t^2 - 4)\cos(9t^3 - 4t)$

4. $f(x) = \dfrac{3x^4 - 7x^2}{4x - 3}$

$f'(x) = \dfrac{(4x-3)(12x^3 - 14x) - (3x^4 - 7x^2)(4)}{(4x-3)^2}$

5. $f(b) = e^{3b^2 - 5b + 2}$

$f'(b) = (6b - 5)e^{3b^2 - 5b + 2}$

6. $f(x) = \sqrt{2x + 5}$

$f(x) = (2x + 5)^{\frac{1}{2}}$

$f'(x) = \dfrac{1}{2}(2x + 5)^{-\frac{1}{2}}(2)$

7. $f(x) = \dfrac{5}{x^4} + \cos(4x)$

$f(x) = 5x^{-4} + \cos(4x)$

$f'(x) = -20x^{-5} - 4\sin(4x)$

8. $f(t) = (6t^2 - 4t)(5t^3 + 7t - 3)$

$f'(t) = (6t^2 - 4t)(15t^2 + 7) + (5t^3 + 7t - 3)(12t - 4)$

9. $f(d) = 8d^4 - 6d^3 + 3d$

$f'(d) = 32d^3 - 18d^2 + 3$

10. $f(x) = \dfrac{e^{3x}}{2x - 7}$

$f'(x) = \dfrac{(2x-7)(3e^{3x}) - e^{3x}(2)}{(2x-7)^2}$

11. $f(x) = 9\ln(x^2)$

$f'(x) = 9 \cdot \dfrac{2x}{x^2} = \dfrac{18}{x}$

12. $f(t) = e^{5t}(4t^2 - 11t)$

$f'(t) = e^{5t}(8t - 11) + (4t^2 - 11t)(5e^{5t})$

13. $f(w) = \sin(5w) + \dfrac{4w - 12}{w^2 + 2w}$

$f'(w) = 5\cos(5w) + \dfrac{(w^2 + 2w)(4) - (4w - 12)(2w + 2)}{(w^2 + 2w)^2}$

14. $f(x) = \dfrac{\cos(8x + 2)}{\sqrt{x}}$

$f(x) = \dfrac{\cos(8x + 2)}{x^{\frac{1}{2}}}$

$f'(x) = \dfrac{x^{\frac{1}{2}}[-8\sin(8x + 2)] - \cos(8x + 2)\left(\dfrac{1}{2}x^{-\frac{1}{2}}\right)}{\left(x^{\frac{1}{2}}\right)^2}$

15. $f(x) = \sqrt[3]{9x^2 + 5x - 6}$

$f(x) = (9x^2 + 5x - 6)^{\frac{1}{3}}$

$f'(x) = \dfrac{1}{3}(9x^2 + 5x - 6)^{-\frac{2}{3}}(18x + 5)$

SOLUTIONS FOR DIFFERENTIAL CALCULUS 4: Differentials

pp. 71 - 72

Find the derivatives of the following functions. Express each derivative using differential notation. If necessary, refer to Differential Calculus 3 for a review of the rules for calculating the various types of derivatives.

1. $y = 6x$

    $\dfrac{dy}{dx} = 6$

2. $y = 7x^2 + 3x + 2$

    $\dfrac{dy}{dx} = 14x + 3$

3. $y = e^{4x}$

    $\dfrac{dy}{dx} = 4e^{4x}$

4. $B = 5t^4 + 7t^2$

    $\dfrac{dB}{dt} = 20t^3 + 14t$

5. $E = \sin(5x)$

    $\dfrac{dE}{dx} = 5\cos(5x)$

6. $x = 5\cos(2t)$

    $\dfrac{dx}{dt} = -10\sin(2t)$

7. $q = \cos t$

    $\dfrac{dq}{dt} = -\sin t$

8. $p = (4t^2 + 3t)^5$

    $\dfrac{dp}{dt} = 5(4t^2 + 3t)^4 (8t + 3)$

9. $v = \dfrac{5}{t^3}$

    $v = 5t^{-3}$

    $\dfrac{dv}{dt} = -15t^{-4}$

10. $y = \sqrt{x}$

    $y = x^{\frac{1}{2}}$

    $\dfrac{dy}{dx} = \dfrac{1}{2}x^{-\frac{1}{2}}$

11. $y = \ln x$

$\dfrac{dy}{dx} = \dfrac{1}{x}$

12. $B = \tan t$

$\dfrac{dB}{dt} = \sec^2 t$

13. $y = e^{9x^2} + \ln(x^2 + 7x)$

$\dfrac{dy}{dx} = 18xe^{9x^2} + \dfrac{2x+7}{x^2 + 7x}$

14. $B = (x^3 + 7x)\ln(6x)$

$\dfrac{dB}{dx} = (x^3 + 7x)\left(\dfrac{6}{6x}\right) + \ln(6x)(3x^2 + 7)$

$\dfrac{dB}{dx} = \dfrac{x^3 + 7x}{x} + \ln(6x)(3x^2 + 7)$

$\dfrac{dB}{dx} = x^2 + 7 + \ln(6x)(3x^2 + 7)$

15. $y = \sec x$

$\dfrac{dy}{dx} = \sec x \tan x$

SOLUTIONS FOR DIFFERENTIAL CALCULUS 5: Higher-Order Derivatives

pp. 77 - 78

Section 1

Find the first and second derivatives of the following functions using the notation from Example 1:

1. $f(x) = 5x^6 - 7x^3 + 9x$

    $f'(x) = 30x^5 - 21x^2 + 9$
    $f''(x) = 150x^4 - 42x$

2. $f(x) = 3x^4 + 6x^3$

    $f'(x) = 12x^3 + 18x^2$
    $f''(x) = 36x^2 + 36x$

3. $f(x) = 5x^3 - 9$

    $f'(x) = 15x^2$
    $f''(x) = 30x$

4. $f(x) = \sqrt{x}$

    $f(x) = x^{\frac{1}{2}}$
    $f'(x) = \frac{1}{2}x^{-\frac{1}{2}}$
    $f''(x) = -\frac{1}{4}x^{-\frac{3}{2}}$

5. $f(x) = \dfrac{9}{x^2}$

    $f(x) = 9x^{-2}$
    $f'(x) = -18x^{-3}$
    $f''(x) = 54x^{-4}$

Section 2

Find the first and second derivatives of the following functions using the notation from Example 2:

6. $y = 5x^3 + 6x^2$

$\dfrac{dy}{dx} = 15x^2 + 12x$

$\dfrac{d^2y}{dx^2} = 30x + 12$

7. $y = 3x^{-4}$

$\dfrac{dy}{dx} = -12x^{-5}$

$\dfrac{d^2y}{dx^2} = 60x^{-6}$

8. $y = \dfrac{3}{4}x^5 - \sqrt{x}$

$y = \dfrac{3}{4}x^5 - x^{\frac{1}{2}}$

$\dfrac{dy}{dx} = \dfrac{15}{4}x^4 - \dfrac{1}{2}x^{-\frac{1}{2}}$

$\dfrac{d^2y}{dx^2} = 15x^3 + \dfrac{1}{4}x^{-\frac{3}{2}}$

9. $y = e^{6x}$

$\dfrac{dy}{dx} = 6e^{6x}$

$\dfrac{d^2y}{dx^2} = 36e^{6x}$

10. $y = \sin(2x)$

$\dfrac{dy}{dx} = 2\cos(2x)$

$\dfrac{d^2y}{dx^2} = -4\sin(2x)$

Section 3

Find the first and second derivatives of the following functions. Express each derivative using the appropriate dependent and independent variables: dB/dt, dP/dr, etc.

11. $x = 5t^4 - 8t^3$

$\dfrac{dx}{dt} = 20t^3 - 24t^2$

$\dfrac{d^2x}{dt^2} = 60t^2 - 48t$

12. $F = 7s^3 + 8s - 3$

$\dfrac{dF}{ds} = 21s^2 + 8$

$\dfrac{d^2F}{ds^2} = 42s$

13. $P = 4w^6 - \dfrac{2}{3}w^4$

$\dfrac{dP}{dw} = 24w^5 - \dfrac{8}{3}w^3$

$\dfrac{d^2P}{dw^2} = 120w^4 - 8w^2$

14. $B = \sqrt{t}$

$B = t^{\frac{1}{2}}$

$\dfrac{dB}{dt} = \dfrac{1}{2}t^{-\frac{1}{2}}$

$\dfrac{d^2B}{dt^2} = -\dfrac{1}{4}t^{-\frac{3}{2}}$

15. $Z = \dfrac{4}{g^3} + 7g$

$z = 4g^{-3} + 7g$

$\dfrac{dz}{dg} = -12g^{-4} + 7$

$\dfrac{d^2z}{dg^2} = 48g^{-5}$

Solutions 211

SOLUTIONS FOR DIFFERENTIAL CALCULUS 6:
Implicit Differentiation

p. 85

Implicitly differentiate the following expressions:

1. $y^2 = 6x$

    $2y\dfrac{dy}{dx} = 6$

    $\dfrac{dy}{dx} = \dfrac{6}{2y}$

2. $y^2 + 3x^5 = 4x$

    $2y\dfrac{dy}{dx} + 15x^4 = 4$

    $2y\dfrac{dy}{dx} = 4 - 15x^4$

    $\dfrac{dy}{dx} = \dfrac{4 - 15x^4}{2y}$

3. $y^3 + y^2 = 7x^4 - 6x$

    $3y^2\dfrac{dy}{dx} + 2y\dfrac{dy}{dx} = 28x^3 - 6$

    $\dfrac{dy}{dx}(3y^2 + 2y) = 28x^3 - 6$

    $\dfrac{dy}{dx} = \dfrac{28x^3 - 6}{3y^2 + 2y}$

4. $4y^6 - 3x^2 = 5y^2 + 9x$

    $24y^5\dfrac{dy}{dx} - 6x = 10y\dfrac{dy}{dx} + 9$

    $24y^5\dfrac{dy}{dx} - 10y\dfrac{dy}{dx} = 9 + 6x$

    $\dfrac{dy}{dx}(24y^5 - 10y) = 9 + 6x$

    $\dfrac{dy}{dx} = \dfrac{9 + 6x}{24y^5 - 10y}$

5. $3y^4 + \sin(5x) = 7$

    $12y^3\dfrac{dy}{dx} + 5\cos(5x) = 0$

    $12y^3\dfrac{dy}{dx} = -5\cos(5x)$

    $\dfrac{dy}{dx} = \dfrac{-5\cos(5x)}{12y^3}$

Implicitly differentiate the following functions using the designated dependent and independent variables:

6. $i^2 = 5t$

$$2i\frac{di}{dt} = 5$$

$$\frac{di}{dt} = \frac{5}{2i}$$

7. $q^3 + 2t = 4t^3$

$$3q^2\frac{dq}{dt} + 2 = 12t^2$$

$$3q^2\frac{dq}{dt} = 12t^2 - 2$$

$$\frac{dq}{dt} = \frac{12t^2 - 2}{3q^2}$$

8. $x^3 = e^{9t}$

$$3x^2\frac{dx}{dt} = 9e^{9t}$$

$$\frac{dx}{dt} = \frac{9e^{9t}}{3x^2}$$

9. $p^2 + \cos(2x) = 6x^3$

$$2p\frac{dp}{dx} - 2\sin(2x) = 18x^2$$

$$2p\frac{dp}{dx} = 18x^2 + 2\sin(2x)$$

$$\frac{dp}{dx} = \frac{18x^2 + 2\sin(2x)}{2p}$$

10. $D^4 + \sqrt{R} = \dfrac{8}{R^2}$

$$D^4 + R^{\frac{1}{2}} = 8R^{-2}$$

$$4D^3\frac{dD}{dR} + \frac{1}{2}R^{-\frac{1}{2}} = -16R^{-3}$$

$$4D^3\frac{dD}{dR} = -16R^{-3} - \frac{1}{2}R^{-\frac{1}{2}}$$

$$\frac{dD}{dR} = \frac{-16R^{-3} - \frac{1}{2}R^{-\frac{1}{2}}}{4D^3}$$

Solutions for Differential Calculus 7: Maxima, Minima, and Points of Inflection

p. 100

Find any/all local maxima, local minima, and points of inflection for the following functions.

1. $f(x) = 3x^2$

    $f'(x) = 6x$
    $6x = 0$
    $x = 0$
    $f''(x) = 6 > 0$

    A minimum value is occurring at $x = 0$.

2. $f(x) = -x^2$

    $f'(x) = -2x$
    $-2x = 0$
    $x = 0$
    $f''(x) = -2 < 0$

    A maximum value is occurring at $x = 0$.

3. $f(x) = \dfrac{1}{3}x^3 - \dfrac{1}{2}x^2 - 12x + 6$

    $f'(x) = x^2 - x - 12$
    $x^2 - x - 12 = 0$
    $(x-4)(x+3) = 0$

    $x - 4 = 0 \qquad\qquad x + 3 = 0$
    $x = 4 \qquad\qquad\quad\; x = -3$

$f''(x) = 2x - 1$

$f''(4) = 2(4) - 1 = 7 > 0$

A minimum value is occurring at $x = 4$.

$f''(-3) = 2(-3) - 1 = -7 < 0$

A maximum value is occurring at $x = -3$.

4. $f(x) = 8x^2 - 1$

$f'(x) = 16x$
$16x = 0$
$x = 0$
$f''(x) = 16 > 0$

A minimum value is occurring at $x = 0$.

5. $f(x) = x^3$

$f'(x) = 3x^2$
$3x^2 = 0$
$x = 0$
$f''(x) = 6x$
$f''(0) = 6(0) = 0$

Since the second derivative evaluated at $x = 0$ is zero, a point of inflection may be occurring at $x = 0$. We must now evaluate the second derivative at points to the left and right of $x = 0$.

$f''(-0.1) = 6(-0.1) = -0.6 < 0$
$f''(0.1) = 6(0.1) = 0.6 > 0$

Since the sign of the second derivative is different to the right and to the left of $x = 0$, a point of inflection is occurring at $x = 0$.

SOLUTIONS FOR INTEGRAL CALCULUS 2:
Finding the Integral of a Function

pp. 134 - 139

Antiderivative Exercises

Find the antiderivative of the following functions. Remember to attach the unknown constant, C, to your final answer.

Section 1

1. $f(x) = 2x$

   Antiderivative: $x^2 + C$

2. $f(x) = 3x^2$

   Antiderivative: $x^3 + C$

3. $f(x) = 6x + 4$

   Antiderivative: $3x^2 + 4x + C$

4. $f(x) = 5x^2 + 7x - 2$

   Antiderivative: $\frac{5}{3}x^3 + \frac{7}{2}x^2 - 2x + C$

5. $f(x) = 9x^3 - 3x$

   Antiderivative: $\frac{9}{4}x^4 - \frac{3}{2}x^2 + C$

Section 2

6. $f(x) = \sqrt{x}$

   $f(x) = x^{\frac{1}{2}}$

   Antiderivative: $\frac{2}{3}x^{\frac{3}{2}} + C$

7. $f(x) = \frac{3}{x^2}$

   $f(x) = 3x^{-2}$

   Antiderivative: $-3x^{-1} + C$

216 Solutions

8. $f(x) = 8x^2 - \sqrt[3]{x}$

$f(x) = 8x^2 - x^{\frac{1}{3}}$

Antiderivative: $\dfrac{8}{3}x^3 - \dfrac{3}{4}x^{\frac{4}{3}} + C$

9. $f(x) = \dfrac{4}{\sqrt{x}}$

$f(x) = 4x^{-\frac{1}{2}}$

Antiderivative: $8x^{\frac{1}{2}} + C$

10. $f(x) = x^5 - 3x^4 + \sqrt{x}$

$f(x) = x^5 - 3x^4 + x^{\frac{1}{2}}$

Antiderivative: $\dfrac{1}{6}x^6 - \dfrac{3}{5}x^5 + \dfrac{2}{3}x^{\frac{3}{2}} + C$

Section 3

11. $f(x) = \cos(2x)$

Antiderivative: $\dfrac{1}{2}\sin(2x) + C$

12. $f(x) = e^{3x}$

Antiderivative: $\dfrac{1}{3}e^{3x} + C$

13. $f(x) = \dfrac{3x^2 + 2x}{x^3 + x^2 + 7}$

Antiderivative: $\ln(x^3 + x^2 + 7) + C$

14. $f(x) = \sin(5x)$

Antiderivative: $-\dfrac{1}{5}\cos(5x) + C$

15. $f(x) = e^x + 6\sin x$

Antiderivative: $e^x - \cos(6x) + C$

Indefinite Integral Exercises

Integrate the following indefinite integrals. Remember to include the constant of integration, C, in the final answer.

1. $\displaystyle\int x^2 \, dx = \dfrac{1}{3}x^3 + C$

2. $\displaystyle\int (x-5) \, dx = \dfrac{1}{2}x^2 - 5x + C$

Solutions 217

3. $\int(4x^3 - 5x^2 + 7x)dx = x^4 - \frac{5}{3}x^3 + \frac{7}{2}x^2 + C$

4. $\int \frac{1}{3}x^4 dx = \frac{1}{15}x^5 + C$

5. $\int \sqrt{x}\, dx = \int x^{\frac{1}{2}} dx = \frac{2}{3}x^{\frac{3}{2}} + C$

6. $\int \sqrt[4]{x^3}\, dx = \int x^{\frac{3}{4}} dx = \frac{4}{7}x^{\frac{7}{4}} + C$

7. $\int(6x^3 + 4x - 9)dx = \frac{6}{4}x^4 + 2x^2 - 9x + C$

8. $\int \left(\frac{x}{4}\right)dx = \int \frac{1}{4}x\, dx = \frac{1}{8}x^2 + C$

9. $\int(5x + \sqrt{x})dx = \int(5x + x^{\frac{1}{2}})dx = \frac{5}{2}x^2 + \frac{2}{3}x^{\frac{3}{2}} + C$

10. $\int 4\, dx = 4x + C$

11. $\int \sin 2x\, dx = -\frac{1}{2}\cos(2x) + C$

12. $\int e^{8x} dx = \frac{1}{8}e^{8x} + C$

13. $\int(x^5 + \cos x)dx = \frac{1}{6}x^6 + \sin x + C$

14. $\int \frac{1}{x} dx = \ln x + C$

15. $\int(6 - e^x)dx = 6x - e^x + C$

Definite Integral Exercises

Integrate the following definite integrals:

1.
$$\int_0^1 x^2 dx = \left[\frac{1}{3}x^3 + C\right]_0^1 = \left[\frac{1}{3}(1)^3 + C\right] - \left[\frac{1}{3}(0)^3 + C\right]$$

$$\int_0^1 x^2 dx = \frac{1}{3} + C - C$$

$$\int_0^1 x^2 dx = \frac{1}{3}$$

218   Solutions

2.

$$\int_1^3 (x^2 - 2)dx = \left[\frac{1}{3}x^3 - 2x + C\right]_1^3 = \left[\frac{1}{3}(3)^3 - 2(3) + C\right] - \left[\frac{1}{3}(1)^3 - 2(1) + C\right]$$

$$\int_1^3 (x^2 - 2)dx = [9 - 6 + C] - \left[\frac{1}{3} - 2 + C\right]$$

$$\int_1^3 (x^2 - 2)dx = [3 + C] - \left[-\frac{5}{3} + C\right]$$

$$\int_1^3 (x^2 - 2)dx = \frac{14}{3} = 4.67$$

3.

$$\int_1^5 \frac{1}{2}x \, dx = \left[\frac{1}{4}x^2 + C\right]_1^5 = \left[\frac{1}{4}(5)^2 + C\right] - \left[\frac{1}{4}(1)^2 + C\right]$$

$$\int_1^5 \frac{1}{2}x \, dx = \left[\frac{25}{4} + C\right] - \left[\frac{1}{4} + C\right]$$

$$\int_1^5 \frac{1}{2}x \, dx = \frac{24}{4} = 6$$

4.

$$\int_1^2 (3x^2 - x)dx = \left[x^3 - \frac{1}{2}x^2 + C\right]_1^2 = \left[(2)^3 - \frac{1}{2}(2)^2 + C\right] - \left[(1)^3 - \frac{1}{2}(1)^2 + C\right]$$

$$\int_1^2 (3x^2 - x)dx = [8 - 2 + C] - \left[1 - \frac{1}{2} + C\right]$$

$$\int_1^2 (3x^2 - x)dx = 5\frac{1}{2} = 5.5$$

5.
$$\int_0^2 \sqrt{x}\, dx = \int_0^2 x^{\frac{1}{2}} dx = \left[\frac{2}{3}x^{\frac{3}{2}} + C\right] = \left[\frac{2}{3}(2)^{\frac{3}{2}} + C\right] - \left[\frac{2}{3}(0)^{\frac{3}{2}} + C\right]$$

$$\int_0^2 \sqrt{x}\, dx = 1.89$$

6.
$$\int_2^4 \frac{5}{x^2} dx = \int_2^4 5x^{-2} dx = \left[-5x^{-1} + C\right]_2^4 = \left[\frac{-5}{x} + C\right]_2^4$$

$$\int_2^4 \frac{5}{x^2} dx = \left[\frac{-5}{4} + C\right] - \left[\frac{-5}{2} + C\right]$$

$$\int_2^4 \frac{5}{x^2} dx = \frac{5}{4} = 1.25$$

7.
$$\int_{1.5}^{2.3} x^2 dx = \left[\frac{1}{3}x^3 + C\right]_{1.5}^{2.3} = \left[\frac{1}{3}(2.3)^3 + C\right] - \left[\frac{1}{3}(1.5)^3 + C\right]$$

$$\int_{1.5}^{2.3} x^2 dx = [1.76 + C] - [0.75 + C] = 1.01$$

8.
$$\int_0^{0.7} (3x - 2) dx = \left[\frac{3}{2}x^2 - 2x + C\right]_0^{0.7} = \left[\frac{3}{2}(0.7)^2 - 2(0.7) + C\right] - \left[\frac{3}{2}(0)^2 - 2(0) + C\right]$$

$$\int_0^{0.7} (3x - 2) dx = [0.735 - 1.4 + C] - [0 + C] = -0.665$$

9.

$$\int_1^3 \frac{\sqrt[3]{x^2}}{4}dx = \int_1^3 \frac{1}{4}x^{\frac{2}{3}}dx = \left[\frac{3}{20}x^{\frac{5}{3}}+C\right]_1^3 = \left[\frac{3}{20}(3)^{\frac{5}{3}}+C\right]-\left[\frac{3}{20}(1)^{\frac{5}{3}}+C\right]$$

$$\int_1^3 \frac{\sqrt[3]{x^2}}{4}dx = [0.94+C]-[0.15+C] = 0.79$$

10.

$$\int_2^3 (x+4)dx = \left[\frac{1}{2}x^2+4x+C\right]_2^3 = \left[\frac{1}{2}(3)^2+4(3)+C\right]-\left[\frac{1}{2}(2)^2+4(2)+C\right]$$

$$\int_2^3 (x+4)dx = [4.5+12+C]-[2+8+C] = 6.5$$

11.

$$\int_0^1 e^x dx = [e^x+C]_0^1 = [e^1+C]-[e^0+C]$$

$$\int_0^1 e^x dx = 1.71828$$

12.

$$\int_0^{\frac{\pi}{4}} \cos(2x) = \left[\frac{1}{2}\sin(2x)\right]_0^{\frac{\pi}{4}} = \left[\frac{1}{2}\sin\left(2\cdot\frac{\pi}{4}\right)+C\right]-\left[\frac{1}{2}\sin(2\cdot 0)+C\right]$$

$$\int_0^{\frac{\pi}{4}} \cos(2x) = \left[\frac{1}{2}\sin\frac{\pi}{2}+C\right]-\left[\frac{1}{2}\sin 0+C\right]$$

$$\int_0^{\frac{\pi}{4}} \cos(2x) = \left[\frac{1}{2}\cdot 1+C\right]-\left[\frac{1}{2}\cdot 0+C\right] = \frac{1}{2}$$

13.
$$\int_1^2 \frac{2x+3}{x^2+3x}dx = \left[\ln(x^2+3x)+C\right]_1^2 = \left[\ln(2^2+3(2))+C\right] - \left[\ln(1^2+3(1))+C\right]$$

$$\int_1^2 \frac{2x+3}{x^2+3x}dx = \left[\ln(10)+C\right] - \left[\ln(4)+C\right] = 0.916$$

14.
$$\int_1^5 dx = \left[x+C\right]_1^5 = \left[5+C\right] - \left[1+C\right]$$

$$\int_1^5 dx = 4$$

15.
$$\int_0^{\frac{\pi}{12}} 12\sin 6x\, dx = \left[-2\cos(6x)+C\right]_0^{\frac{\pi}{12}} = \left[-2\cos\left(6\cdot\frac{\pi}{12}\right)+C\right] - \left[-2\cos(6\cdot 0)+C\right]$$

$$\int_0^{\frac{\pi}{12}} 12\sin 6x\, dx = \left[-2\cos\left(\frac{\pi}{2}\right)+C\right] - \left[-2\cos(0)+C\right]$$

$$\int_0^{\frac{\pi}{12}} 12\sin 6x\, dx = \left[-2(0)+C\right] - \left[-2(1)+C\right] = 2$$

222 Solutions

SOLUTIONS FOR INTEGRAL CALCULUS 3:
The Method of Substitution

pp. 148 - 149

Use the method of substitution to evaluate the following definite and indefinite integrals.

1. $\int (3x^2 + 4)^3 6x\,dx$

   $u = 3x^2 + 4$
   $\dfrac{du}{dx} = 6x$
   $du = 6x\,dx$

   $\int (3x^2 + 4)^3 6x\,dx = \int u^3\,du$

   $\int u^3\,du = \dfrac{1}{4}u^4 + C$

   Reinserting the function of x:

   $\int (3x^2 + 4)^3 6x\,dx = \dfrac{1}{4}(3x^2 + 4)^4 + C$

2. $\int (7x^4 - 5x^2)^4 (28x^3 - 10x)\,dx$

   $u = 7x^4 - 5x^2$
   $\dfrac{du}{dx} = 28x^3 - 10x$
   $du = (28x^3 - 10x)\,dx$

$$\int (7x^4 - 5x^2)^4 (28x^3 - 10x) dx = \int u^4 du$$

$$\int u^4 du = \frac{1}{5} u^5 + C$$

Reinserting the function of x:

$$\int (7x^4 - 5x^2)^4 (28x^3 - 10x) dx = \frac{1}{5}(7x^4 - 5x^2)^5 + C$$

3. $\int (6x^3 + 5x + 4)^6 (18x^2 + 5) dx$

$u = 6x^3 + 5x + 4$

$\dfrac{du}{dx} = 18x^2 + 5$

$du = (18x^2 + 5) dx$

$$\int (6x^3 + 5x + 4)^6 (18x^2 + 5) dx = \int u^6 du$$

$$\int u^6 du = \frac{1}{7} u^7 + C$$

Reinserting the function of x:

$$\int (6x^3 + 5x + 4)^6 (18x^2 + 5) dx = \frac{1}{7}(6x^3 + 5x + 4)^7 + C$$

4. $\int (2x^3 + 3x)^4 (12x^2 + 6) dx$

$u = 2x^3 + 3x$

$\dfrac{du}{dx} = 6x^2 + 3$

$du = (6x^2 + 3) dx$

224   Solutions

Inserting a factor of ½ inside the integral and a factor of 2 in front of the integral:

$$\int (2x^3+3x)^4(12x^2+6)dx = 2\int(2x^3+3x)^4 \frac{1}{2}(12x^2+6)dx$$

$$\int (2x^3+3x)^4(12x^2+6)dx = 2\int(2x^3+3x)^4(6x^2+3)dx$$

$$2\int (2x^3+3x)^4(6x^2+3)dx = 2\int u^4 du$$

$$2\int u^4 du = 2\left(\frac{1}{5}u^5+C\right) = \frac{2}{5}u^5+C$$

Reinserting the function of x:

$$\int(2x^3+3x)^4(12x^2+6)dx = \frac{2}{5}(2x^3+3x)^5+C$$

5. $$\int_0^1 (x^2+x)^2(2x+1)dx$$

$u = x^2+x$
$\dfrac{du}{dx} = 2x+1$
$du = (2x+1)dx$

$$\int(x^2+x)^2(2x+1)dx = \int u^2 du$$

$$\int u^2 du = \frac{1}{3}u^3+C$$

Reinserting the function of x:

$$\int_0^1(x^2+x)^2(2x+1)dx = \left[\frac{1}{3}(x^2+x)^3+C\right]_0^1$$

Inserting the endpoints:

$$\int_0^1 (x^2+x)^2(2x+1)dx = \left[\frac{1}{3}(1^2+1)^3 + C\right] - \left[\frac{1}{3}(0^2+0)^3 + C\right]$$

$$\int_0^1 (x^2+x)^2(2x+1)dx = \frac{8}{3}$$

6. $\int (2x^7 - 8x^3)^5 (14x^6 - 24x^2)dx$

$u = 2x^7 - 8x^3$
$\dfrac{du}{dx} = 14x^6 - 24x^2$
$du = (14x^6 - 24x^2)dx$

$$\int (2x^7 - 8x^3)^5 (14x^6 - 24x^2)dx = \int u^5 du$$

$$\int u^5 du = \frac{1}{6}u^6 + C$$

Reinserting the function of x:

$$\int (2x^7 - 8x^3)^5 (14x^6 - 24x^2)dx = \frac{1}{6}(2x^7 - 8x^3)^6 + C$$

7. $\int (4x^2 + 6x)^5 (4x+3)dx$

$u = 4x^2 + 6x$
$\dfrac{du}{dx} = 8x + 6$
$du = (8x+6)dx$

226   Solutions

Inserting a factor of 2 inside the integral and a factor of ½ in front of the integral:

$$\int (4x^2+6x)^5 (4x+3)dx = \frac{1}{2}\int (4x^2+6x)^5 2(4x+3)dx$$

$$\int (4x^2+6x)^5 (4x+3)dx = \frac{1}{2}\int (4x^2+6x)^5 (8x+6)dx$$

$$\frac{1}{2}\int (4x^2+6x)^5 (8x+6)dx = \frac{1}{2}\int u^5 du$$

$$\frac{1}{2}\int u^5 du = \frac{1}{2}\left(\frac{1}{6}u^6 + C\right) = \frac{1}{12}u^6 + C$$

Reinserting the function of x:

$$\int (4x^2+6x)^5 (4x+3)dx = \frac{1}{12}(4x^2+6x)^6 + C$$

8. $\int_{2}^{3} 6(6x-1)^4 dx$

$$\int_{2}^{3} 6(6x-1)^4 dx = \int_{2}^{3} (6x-1)^4 6dx$$

$u = 6x-1$
$\dfrac{du}{dx} = 6$
$du = 6dx$

$$\int (6x-1)^4 6dx = \int u^4 du$$

$$\int u^4 du = \frac{1}{5}u^5 + C$$

Reinserting the function of x:

$$\int_2^3 6(6x-1)^4\,dx = \left[\frac{1}{5}(6x-1)^5 + C\right]_2^3$$

Inserting the endpoints:

$$\int_2^3 6(6x-1)^4\,dx = \left[\frac{1}{5}(6(3)-1)^5 + C\right] - \left[\frac{1}{5}(6(2)-1)^5 + C\right]$$

$$\int_2^3 6(6x-1)^4\,dx = [283971.4 + C] - [32210.2 + C]$$

$$\int_2^3 6(6x-1)^4\,dx = 251761.2$$

9. $$\int_0^2 (8x^2 + 4x + 3)^3 (4x+1)\,dx$$

$u = 8x^2 + 4x + 3$
$\dfrac{du}{dx} = 16x + 4$
$du = (16x+4)\,dx$

Inserting a factor of 4 inside the integral and a factor of ¼ in front of the integral:

$$\int_0^2 (8x^2+4x+3)^3(4x+1)\,dx = \frac{1}{4}\int_0^2 (8x^2+4x+3)^3 \, 4(4x+1)\,dx$$

$$\int_0^2 (8x^2+4x+3)^3(4x+1)\,dx = \frac{1}{4}\int_0^2 (8x^2+4x+3)^3 (16x+4)\,dx$$

$$\frac{1}{4}\int(8x^2+4x+3)^3(16+4)dx = \frac{1}{4}\int u^3 du$$

$$\frac{1}{4}\int u^3 du = \frac{1}{4}\left(\frac{1}{4}u^4+C\right) = \frac{1}{16}u^4+C$$

Reinserting the function of x:

$$\int_0^2 (8x^2+4x+3)^3(4x+1)dx = \left[\frac{1}{16}(8x^2+4x+3)^4+C\right]_0^2$$

Inserting the endpoints:

$$\int_0^2 (8x^2+4x+3)^3(4x+1)dx = \left[\frac{1}{16}(8(2)^2+4(2)+3)^4+C\right]-\left[\frac{1}{16}(8(0)^2+4(0)+3)^4+C\right]$$

$$\int_0^2 (8x^2+4x+3)^3(4x+1)dx = [213675.1+C]-[5.1+C]$$

$$\int_0^2 (8x^2+4x+3)^3(4x+1)dx = 213670$$

10. $\int\dfrac{(\sqrt{x}+3)^5}{2\sqrt{x}}dx$

$$\int\frac{(\sqrt{x}+3)^5}{2\sqrt{x}}dx = \int(x^{\frac{1}{2}}+3)^5\frac{1}{2}x^{-\frac{1}{2}}dx$$

$u = x^{\frac{1}{2}}+3$

$\dfrac{du}{dx} = \dfrac{1}{2}x^{-\frac{1}{2}}$

$du = \left(\dfrac{1}{2}x^{-\frac{1}{2}}\right)dx$

$$\int \left(x^{\frac{1}{2}}+3\right)^5 \frac{1}{2}x^{-\frac{1}{2}}dx = \int u^5 du$$

$$\int u^5 du = \frac{1}{6}u^6 + C$$

Reinserting the function of x:

$$\int \frac{\left(\sqrt{x}+3\right)^5}{2\sqrt{x}} dx = \frac{1}{6}\left(\sqrt{x}+3\right)^6 + C$$

Solutions for Integral Calculus 4:
Differentials II – The Relationship between Integral and Differential Calculus

p. 156

Find the differential for each of the following functions.

1. $y = 5x^4$

   $\dfrac{dy}{dx} = 20x^3$

   $dy = 20x^3 dx$

2. $y = e^{3x}$

   $\dfrac{dy}{dx} = 3e^{3x}$

   $dy = 3e^{3x} dx$

3. $B = \cos 5x$

   $\dfrac{dB}{dx} = -5\sin 5x$

   $dB = -5\sin 5x\, dx$

4. $y = (2t^3 + 7t)^3$

   $\dfrac{dy}{dt} = 3(2t^3 + 7t)^2 (6t^2 + 7)$

   $dy = 3(2t^3 + 7t)^2 (6t^2 + 7)\, dt$

5. $i = \tan t + \sqrt{t}$

   $\dfrac{di}{dt} = \sec^2 t + \dfrac{1}{2}t^{-\frac{1}{2}}$

   $di = \left(\sec^2 t + \dfrac{1}{2}t^{-\frac{1}{2}}\right) dt$

Use differentials to find the change in the dependent variable as the independent variable is changed between the given limits.

6. $y = 2x^2;$ $x = 0$ to $x = 1$

$$\frac{dy}{dx} = 4x$$
$$dy = 4x\,dx$$
$$y = \int_0^1 4x\,dx$$
$$y = \left[2x^2 + C\right]_0^1 = \left[2(1)^2 + C\right] - \left[2(0)^2 + C\right] = 2$$

7. $y = e^x;$ $x = 1$ to $x = 2$

$$\frac{dy}{dx} = e^x$$
$$dy = e^x\,dx$$
$$y = \int_1^2 e^x\,dx$$
$$y = \left[e^2 + C\right] - \left[e^1 + C\right] = 4.67$$

8. $q = (t+1)^3;$ $t = 0$ to $t = 3$

$$\frac{dq}{dt} = 3(t+1)^2$$
$$dq = 3(t+1)^2\,dt$$
$$q = \int_0^3 3(t+1)^2\,dt$$
$$q = \left[(t+1)^3 + C\right]_0^3 = \left[(3+1)^3 + C\right] - \left[(0+1)^3 + C\right] = 63$$

9. $\dfrac{dB}{dt} = e^{-2t}$

$$\dfrac{dB}{dt} = e^{-2t}$$
$$dB = e^{-2t} dt$$
$$B = \int_0^1 e^{-2t} dt$$
$$B = \left[-\dfrac{1}{2}e^{-2t} + C\right]_0^1 = \left[-\dfrac{1}{2}e^{-2(1)} + C\right] - \left[-\dfrac{1}{2}e^{-2(0)} + C\right] = 0.568$$

10. $y = 4.9t^2$

$$\dfrac{dy}{dt} = 9.8t$$
$$dy = 9.8t\, dt$$
$$y = \int_1^3 9.8t\, dt$$
$$y = \left[4.9t^2 + C\right]_1^3 = \left[4.9(3)^2 + C\right] - \left[4.9(1)^2 + C\right] = 39.2$$

Solutions for Integral Calculus 5: Integration by Parts

p. 168

Use the method of Integration by Parts to calculate the following definite and indefinite integrals.

1. $\int xe^{4x}dx$

   $u = x$ $\qquad\qquad dv = e^{4x}dx$

   $\dfrac{du}{dx} = 1$ $\qquad\qquad \dfrac{dv}{dx} = e^{4x}$

   $du = dx$ $\qquad\qquad v = \dfrac{1}{4}e^{4x}$

   $\int u\,dv = uv - \int v\,du$

   $\int xe^{4x}dx = x\left(\dfrac{1}{4}e^{4x}\right) - \int \left(\dfrac{1}{4}e^{4x}\right)dx$

   $\int xe^{4x}dx = \dfrac{1}{4}xe^{4x} - \dfrac{1}{16}e^{4x} + C$

2. $\int x\sin x\,dx$

   $u = x$ $\qquad\qquad dv = \sin x\,dx$

   $\dfrac{du}{dx} = 1$ $\qquad\qquad \dfrac{dv}{dx} = \sin x$

   $du = dx$ $\qquad\qquad v = -\cos x$

   $\int u\,dv = uv - \int v\,du$

   $\int x\sin x\,dx = x(-\cos x) - \int (-\cos x)dx$

   $\int x\sin x\,dx = -x\cos x + \int \cos x\,dx$

   $\int x\sin x\,dx = -x\cos x + \sin x + C$

3. $$\int_1^2 xe^x dx$$

$u = x$ $\qquad$ $dv = e^x dx$
$\dfrac{du}{dx} = 1$ $\qquad$ $\dfrac{dv}{dx} = e^x$
$du = dx$ $\qquad$ $v = e^x$

$\int u\, dv = uv - \int v\, du$

$\int_1^2 xe^x dx = \left[xe^x\right]_1^2 - \int_1^2 e^x dx$

$\int_1^2 xe^x dx = \left[xe^x\right]_1^2 - \left[e^x\right]_1^2$

$\int_1^2 xe^x dx = \left[2 \cdot e^2 - 1 \cdot e^1\right] - \left[e^2 - e^1\right] = 7.4$

4. $$\int t \cos 2t\, dt$$

$u = t$ $\qquad$ $dv = \cos 2t\, dt$
$\dfrac{du}{dt} = 1$ $\qquad$ $\dfrac{dv}{dt} = \cos 2t$
$du = dt$ $\qquad$ $v = \dfrac{1}{2}\sin 2t$

$\int u\, dv = uv - \int v\, du$

$\int t \cos 2t = t\left(\dfrac{1}{2}\sin 2t\right) - \int\left(\dfrac{1}{2}\sin 2t\right) dt$

$\int t \cos 2t = \dfrac{1}{2} t \sin 2t + \dfrac{1}{4}\cos 2t + C$

5. $$\int x^2 \cos x\, dx$$

$u = x^2$ $\quad\quad\quad\quad dv = \cos x dx$

$\dfrac{du}{dx} = 2x$ $\quad\quad\quad\quad \dfrac{dv}{dx} = \cos x$

$du = 2x dx$ $\quad\quad\quad\quad v = \sin x$

$$\int u dv = uv - \int v du$$

$$\int x^2 \cos x dx = x^2 \sin x - \int (\sin x)(2x dx)$$

$$\int x^2 \cos dx = x^2 \sin x - 2 \int x \sin x dx$$

Using Integration by Parts to complete the integral on the right-hand side:

$u = x$ $\quad\quad\quad\quad dv = \sin x dx$

$\dfrac{du}{dx} = 1$ $\quad\quad\quad\quad \dfrac{dv}{dx} = \sin x$

$du = dx$ $\quad\quad\quad\quad v = -\cos x$

$$\int x^2 \cos x dx = x^2 \sin x - 2 \int x \sin x dx$$

$$\int x^2 \cos x dx = x^2 \sin x - 2 \left[x(-\cos x) - \int (-\cos x) dx \right]$$

$$\int x^2 \cos x dx = x^2 \sin x - 2 \left[-x \cos x + \int \cos x dx \right]$$

$$\int x^2 \cos x dx = x^2 \sin x - 2 \left[-x \cos x + \sin x \right]$$

$$\int x^2 \cos x dx = x^2 \sin x + 2x \cos x - 2 \sin x + C$$

Solutions for Integral Calculus 6: Numerical Integration Techniques

pp. 179 - 180

Using the left-endpoint, right end-point, and trapezoidal approximation techniques, approximate the value of each of the following integrals.

1. $\int_0^1 2x^2 \, dx \qquad n = 5$

$\Delta x = \dfrac{1-0}{5} = 0.2$

Left-endpoint Approximation:

Rectangle	Location of Left End	Height	Area
1	0	0	0
2	0.2	0.08	0.016
3	0.4	0.32	0.064
4	0.6	0.72	0.144
5	0.8	1.28	0.256

Area = 0.48

Right-endpoint Approximation:

Rectangle	Location of Right End	Height	Area
1	0.2	0.08	0.016
2	0.4	0.32	0.064
3	0.6	0.72	0.144
4	0.8	1.28	0.256
5	1	2	0.4

Area = 0.88

Trapezoidal Approximation:

$$\text{Area} = \frac{0.2}{2}[f(0) + 2f(0.2) + 2f(0.4) + 2f(0.6) + 2f(0.8) + f(1)]$$
$$\text{Area} = 0.1[0 + 0.16 + 0.64 + 1.44 + 2.56 + 2]$$
$$\text{Area} = 0.68$$

2. $\int_{1}^{2} e^x dx \qquad n = 10$

$$\Delta x = \frac{2-1}{10} = 0.1$$

Left-endpoint Approximation:

Rectangle	Location of Left End	Height	Area
1	1	2.718	0.2718
2	1.1	3.004	0.3004
3	1.2	3.320	0.3320
4	1.3	3.669	0.3669
5	1.4	4.055	0.4055
6	1.5	4.482	0.4482
7	1.6	4.953	0.4953
8	1.7	5.474	0.5474
9	1.8	6.049	0.6049
10	1.9	6.686	0.6686

Area = 4.441

Right-endpoint Approximation:

Rectangle	Location of Right End	Height	Area
1	1.1	3.004	0.3004
2	1.2	3.320	0.3320
3	1.3	3.669	0.3669
4	1.4	4.055	0.4055
5	1.5	4.482	0.4482
6	1.6	4.953	0.4953
7	1.7	5.474	0.5474
8	1.8	6.049	0.6049
9	1.9	6.686	0.6686
10	2	7.389	0.7389

Area = 4.9081

Trapezoidal Approximation:

$$\text{Area} = \frac{0.1}{2}[f(1)+2f(1.1)+2f(1.2)+2f(1.3)+2f(1.4)+2f(1.5)+2f(1.6)+2f(1.7)+2f(1.8)+2f(1.9)+f(2)]$$

$\text{Area} = 0.05[2.718+6.008+6.640+7.339+8.110+8.963+9.906+10.948+12.099+13.372+7.389]$
$\text{Area} = 4.6746$

3. $\int_{1}^{2}(x^2+3)dx \qquad n=4$

$\Delta x = \frac{2-1}{4} = 0.25$

Left-endpoint Approximation:

Rectangle	Location of Left End	Height	Area
1	1.00	4.000	1.0000
2	1.25	4.562	1.1405
3	1.50	5.250	1.3125
4	1.75	6.062	1.5155

Area = 4.9685

Right-endpoint Approximation:

Rectangle	Location of Right End	Height	Area
1	1.25	4.562	1.1405
2	1.50	5.250	1.3125
3	1.75	6.062	1.5155
4	2.00	7.00	1.75

Area = 5.7185

Trapezoidal Approximation

$$\text{Area} = \frac{0.25}{2}[f(0)+2f(1.25)+2f(1.5)+2f(1.75)+f(2)]$$
$\text{Area} = 0.125[4+9.124+10.500+12.124+7]$
$\text{Area} = 5.3435$

4. $\int_1^4 \frac{1}{x^3} dx \qquad n = 8$

$\Delta x = \frac{4-1}{8} = 0.375$

Left-endpoint Approximation:

Rectangle	Location of Left End	Height	Area
1	1.000	1.000	0.3750
2	1.375	0.385	0.1444
3	1.750	0.187	0.0701
4	2.125	0.104	0.0390
5	2.500	0.064	0.0240
6	2.875	0.042	0.0158
7	3.250	0.029	0.0109
8	3.625	0.021	0.0079

Area = 0.6871

Right-endpoint Approximation:

Rectangle	Location of Right End	Height	Area
1	1.375	0.385	0.1444
2	1.750	0.187	0.0701
3	2.125	0.104	0.0390
4	2.500	0.064	0.0240
5	2.875	0.042	0.0158
6	3.250	0.029	0.0109
7	3.625	0.021	0.0079
8	4	0.016	0.0060

Area = 0.3181

Trapezoidal Approximation:

Area $= \frac{0.375}{2}[f(1) + 2f(1.375) + 2f(1.750) + 2f(2.125) + 2f(2.500) + 2f(2.875) + 2f(3.250) + 2f(3.625) + f(4)]$

Area $= 0.1875[1 + 0.770 + 0.374 + 0.208 + 0.128 + 0.084 + 0.058 + 0.042 + 0.016]$

Area $= 0.5026$

5. $$\int_1^4 \sqrt{x}\, dx \qquad n=6$$

$$\Delta x = \frac{4-1}{6} = 0.5$$

Left-endpoint Approximation:

Rectangle	Location of Left End	Height	Area
1	1.0	1.000	0.5000
2	1.5	1.225	0.6125
3	2.0	1.414	0.7070
4	2.5	1.581	0.7905
5	3.0	1.732	0.8660
6	3.5	1.871	0.9355

Area = 4.4115

Right-endpoint Approximation:

Rectangle	Location of Right End	Height	Area
1	1.5	1.225	0.6125
2	2.0	1.414	0.7070
3	2.5	1.581	0.7905
4	3.0	1.732	0.8660
5	3.5	1.871	0.9355
6	4.0	2.000	1.0000

Area = 4.9115

Trapezoidal Approximation:

$$\text{Area} = \frac{0.5}{2}[f(1) + 2f(1.5) + 2f(2.0) + 2f(2.5) + 2f(3.0) + 2f(3.5) + f(4.0)]$$
$$\text{Area} = 0.25[1.000 + 2.450 + 2.828 + 3.162 + 3.464 + 3.742 + 2]$$
$$\text{Area} = 4.6615$$

Index

Absolute maxima, 95
Absolute minima, 95
Acceleration, 103
Angular acceleration, 108
Angular velocity, 108
Antiderivative of a function, 123
Average value, 183
Capacitor, 105,107
Concavity, 97
Definite integral, 133
Dependent variables, 12
Derivative of a function, 24,29
Derivatives
    Cosecant function, 59
    Cosine function, 47
    Cotangent function, 59
    Exponential function, 48
    Natural logarithm, 49
    Power Rule, 59
    Product Rule, 54
    Quotient Rule, 55
    Secant function, 58
    Simple polynomials, 38
    Sine function, 46
    Tangent function, 57
Differentials, 67, 151
Displacement, 103,183
Electric charge, 104, 107, 154, 182
Electric current, 104, 154, 182
First derivatives, 75
First derivative test, 88
Force, 103, 181, 183
Functional notation, 14
Heat, 186
Higher-order derivatives, 75
Implicit differentiation, 81
Impulse, 182
Indefinite integral, 133
Independent variables, 12
Inductor, 107
Integral of a function, 117
Integration by parts, 159
LC-circuit, 107
Left-endpoint approximation, 172
Limits, 17
Limits of integration, 129

Magnetic field, 109
Maxima, 87
Method of substitution, 143
Minima, 87
Moment of inertia, 185
Momentum, 104
Permeability, 110
PN-junction, 110
Points of inflection, 96
Population growth, 109
Potential energy, 103,183
RC-circuit, 105
Resistor, 105
Right-endpoint approximation, 175
Root-mean-square, 183
Second derivatives, 75
Second derivative test, 91
Shockley's Equation, 110
Simple harmonic oscillators, 104
Slope, 3
Specific heat capacity, 186
Torque, 184
Torricelli's Equation, 111
Trapezoidal approximation, 176
Velocity, 103,183
Voltage, 110
Work, 181, 18